Theory of Quantum and Classical Connections in Modeling Atomic, Molecular and Electrodynamic Systems

Theory of Quantum and Classical Connections in Modeling Atomic, Molecular and Electrodynamic Systems

Alexandru Popa

AMSTERDAM • BOSTON • HEIDELBERG • LONDON
NEW YORK • OXFORD • PARIS • SAN DIEGO
SAN FRANCISCO • SINGAPORE • SYDNEY • TOKYO
Academic Press is an imprint of Elsevier

Academic Press is an imprint of Elsevier
The Boulevard, Langford Lane, Kidlington, Oxford OX5 1GB, UK
Radarweg 29, PO BOX 211, 1000 AE Amsterdam, The Netherlands
225 Wyman Street, Waltham, MA 02451, USA

First published 2014

British Library Cataloguing-in-Publication Data
A catalogue record for this book is available from the British Library

Library of Congress Cataloging-in-Publication Data
A catalog record for this book is available from the Library of Congress

ISBN: 978-0-12-409502-1

For information on all Academic Press publications
visit our website at store.elsevier.com

This book has been manufactured using Print On Demand technology. Each copy
is produced to order and is limited to black ink. The online version of this book
will show color figures where appropriate.

ELSEVIER Book Aid International **Working together to grow libraries in developing countries**

www.elsevier.com • www.bookaid.org

CONTENTS

The basic equations describing the atomic molecular and electrodynamic systems in quantum mechanics are exactly applicable only for very simple cases, so that modern computational physics is a science of the most accurate approximations of the solutions of these equations. For example, the central field approximation leads to the atomic and molecular orbital models for stationary systems.

Numerous analysis models, called semiclassical, are based on relations between quantum and classical equations, which lead to solutions that are relatively easy to calculate, using the information obtained from the study of classical trajectories. Most of the semiclassical models are based on the connection at limit between the Schrödinger and the Hamilton–Jacobi equations, when the value of Planck constant is negligible compared to the action of the system. Another group of models, which do not neglect the value of Planck constant, are based on the hidden variable theory of Bohm, but in this case the Schrödinger equation is related to a modified form of the Hamilton–Jacobi equation, which contains a supplementary term, called quantum potential.

This book is a review of our theoretical works, which is only available in journal articles, where we proved an exact connection between quantum and classical equations. This connection is proved starting from basic equations, without using any approximation. Though our approach is different, there are similarities with the works of Courant, Cox, Synge, and Luis, in the case of stationary multidimensional systems, and with the works of Motz and Selzer in the case of the systems composed of electrons and electrodynamic field.

For stationary multidimensional atomic and molecular systems, the connection between quantum and classical equations is due to the fact that the geometric elements of the wave described by the Schrödinger equation, namely the wave surfaces and their normals, are given by the Hamilton–Jacobi equation, which does not contain the quantum potential term. On the other hand, since the theory is exact, and approximations are made only in applications, our numerical results are in good agreement with the experimental data from literature. In the case of the atomic and molecular stationary systems, we elaborated a calculation method for energetic values, which is based on the wave properties of the system, and whose accuracy is comparable to the accuracy of the Hartree−Fock method.

For electrodynamic systems composed of particle and electromagnetic field, the connection results from the fact that the Klein−Gordon equation is verified exactly by the wave function, corresponding to the classical solution which results from the relativistic Hamilton−Jacobi equation. This connection leads to an accurate method to study the new systems composed from very intense laser fields and electrons or atoms. We present a series of applications of this method, as the calculation of angular and spectral distributions of the radiations generated at interactions between very intense laser beams and electrons or atoms, which are in good agreement with experimental data from literature. The study of these systems is important in this moment, due to the emergence of the new generation of ultraintense lasers. It is the object of some important European research projects.

In spite of the fact that the two systems, stationary atomic and molecular, and nonstationary electrodynamic, are very different, our approach reveals common properties, as exact connections between quantum and classical equations, periodic properties and the fact that the basic quantum equations lead mathematically, without any approximation, to wave functions which correspond to classical solutions.

In this volume we present in detail the connection between the quantum and classical equations, starting from the basic equations, for atomic and molecular systems and respectively for systems comprising electron and electromagnetic field. In a second volume we will present the applications of this theory to modeling atomic and molecular systems, and to modeling properties of radiations generated at the interaction between very intense laser beams and electrons or atoms. The equations are written in the International System.

CHAPTER 1

Connection Between Schrödinger and Hamilton—Jacobi Equations in the Case of Stationary Atomic and Molecular Systems

Abstract

Our approach is based on the equivalence between the Schrödinger and wave equations, which is valid for stationary atomic and molecular systems. We prove that the characteristic surface of the wave equation, which has the significance of a wave surface, and its normal curves are periodic solutions of the Hamilton-Jacobi equation, written for the same system. We prove that the motion of the wave surface is periodic, and that the normal curves are closed. The Bohr generalized relation is valid for these curves, resulting that de Broglie relations are valid for multidimensional systems. We show that the wave surfaces and their normals depend on the same constants of motion as those resulting from the Schrödinger equation. Consequently these geometric elements can be used as analytical tools for calculating the energetic values of the system and for determining its symmetry properties, as detailed in concrete examples in a second book.

Keywords: Schrödinger equation; Hamilton-Jacobi equation; wave equation; wave function; action function; multidimensional systems; stationary systems; atomic systems; molecular systems; wave surface; constants of motion; periodic solutions; integral relations; Bohr generalized relation; de Broglie relations

1.1 INITIAL HYPOTHESES

In this chapter, we review the theoretical results presented in the papers (Popa, 1996, 1998a, b, 1999a, b, c, 2003b, 2005). We shall analyze an atomic or molecular system composed of N electrons and N' nuclei. The Cartesian coordinates of the electrons are x_a, y_a, z_a, where a takes values between 1 and N. Our analysis is made in the space R^{3N} of the electron coordinates, which are denoted by q_j (where $q_1 = x_1$, $q_2 = y_1$, $q_3 = z_1, \ldots, q_{3N} = z_N$), j taking values between 1 and $3N$. We denote by $q = (q_1, q_2, \ldots, q_{3N})$ the coordinate of a point in the space R^{3N}.

We consider the following initial hypotheses: (h1) The system is closed and stationary (i.e., the total energy, denoted by E, is constant and the potential energy, denoted by U, does not depend explicitly on time); (h2) The total energy has real negative values (i.e., the system is in a bound state); (h3) The behavior of the system is completely described by the Schrödinger equation; (h4) The relativistic and magnetic effects are neglected; (h5) The nuclei are fixed on average positions and their motion is neglected.

1.2 SCHÖDINGER EQUATION, WAVE EQUATION, AND CHARACTERISTIC EQUATION

The Schrödinger equation is

$$-i\hbar \frac{\partial \Psi}{\partial t} - \frac{\hbar^2}{2m} \sum_j \frac{\partial^2 \Psi}{\partial q_j^2} + U\Psi = 0 \tag{1.1}$$

where Ψ, m, t, and i are, respectively, the wave function, the electron mass, the time, and the imaginary constant, while $\hbar$ is the normalized Planck constant ($\hbar = h/2\pi$).

The wave function of a stationary system is of the form $\Psi = \Psi(q, t, E, \alpha)$ (see Messiah, 1961, p. 330), E and $\alpha = (\alpha_1, \alpha_2, \ldots, \alpha_S)$ are the eigenvalues of the constants of motion.

Since the system is stationary, the Schrödinger equation has a solution of the form (Landau and Lifschitz, 1991):

$$\Psi = \Psi_0 \exp(-iEt/\hbar) \tag{1.2}$$

where $\Psi_0 = \Psi_0(q, E, \alpha)$ is the time independent wave function, which is a complex valued function that results from the time independent Schrödinger equation:

$$-\frac{\hbar^2}{2m} \sum_j \frac{\partial^2 \Psi_0}{\partial q_j^2} + (U - E)\Psi_0 = 0 \tag{1.3}$$

For stationary systems, Eq. (1.1) is equivalent to the system which comprises Eq. (1.2) and the wave equation

$$\sum_j \frac{\partial^2 \Psi}{\partial q_j^2} - \frac{1}{v_w^2} \frac{\partial^2 \Psi}{\partial t^2} = 0 \qquad (1.4)$$

where

$$v_w = \pm \frac{|E|}{\sqrt{2m(E - U)}} \qquad (1.5)$$

This equivalence is generally valid for stationary systems, because if we introduce Ψ from Eq. (1.2) in Eq. (1.4), we obtain Eq. (1.3). In other words, Eq. (1.1) has rigorously the same solutions as the system of Eqs. (1.2) and (1.4).

The characteristic surface of the wave equation is denoted by Σ, and it is given by the following equation (Courant and Hilbert, 1962; Smirnov, 1984; Zauderer, 1983):

$$\chi(q, t) = 0 \qquad (1.6)$$

where χ is called the *characteristic function*. It is a single valued function which satisfies the *characteristic equation*:

$$\sum_j \left(\frac{\partial \chi}{\partial q_j}\right)^2 - \frac{1}{v_w^2} \left(\frac{\partial \chi}{\partial t}\right)^2 = 0 \qquad (1.7)$$

In virtue of the theory (see Smirnov, 1984, p. 56), an equation of the form $F(\partial u/\partial q_1, \ldots, \partial u/\partial q_M, \partial u/\partial t) = 0$ for which $\partial u/\partial t$ can be written explicitly in terms of the other partial derivatives, has a solution of the form $u = \psi(q, t, c') + c_0'$, where $c' = (c_1', c_2', \ldots, c_M')$ are arbitrary real constants, c_0' is an additive constant, which will be chosen to be zero, and M is the number of independent coordinates. Since one of the constant is the energy E, it follows the complete integral of Eq. (1.7) is of the form

$$\chi = \chi(q, t, E, c) \qquad (1.8)$$

where $c = (c_1, c_2, \ldots, c_{3N-1})$ are the other constants of motion of the system. Since the characteristic surface is an intrinsic mathematical element of the system described by the Schrödinger equation, the c constants are equal to the eigenvalues of the constants of motion which result from the Schrödinger equation. This equality will be detailed in applications (see Section 1.7.1).

The characteristic surface Σ has the significance of a *wave surface* (Courant and Hilbert, 1962; Smirnov, 1984). We analyze the motion of this wave surface in the classically allowed (CA) domain, where $E > U$, corresponding to the real values of v_w given by Eq. (1.5).

1.3 EQUATION OF THE WAVE SURFACES

Equation (1.7) has the following solution:

$$\chi(q, t, E, c) = \sin k[f(q, E, c) \mp |E|t] \tag{1.9}$$

where k is a real constant and $f(q, E, c)$ is a single valued function (the complete integral) which verifies the time independent Hamilton−Jacobi equation

$$\sum_j \left(\frac{\partial f}{\partial q_j}\right)^2 + 2m(U - E) = 0 \tag{1.10}$$

We limit the analysis to the case corresponding to the plus sign in Eq. (1.5) and the minus sign in Eq. (1.9). The case of a minus sign in Eq. (1.5) and a plus sign in Eq. (1.9) corresponds to a set of wave surfaces moving in the opposite direction, as we will show later.

From Eqs. (1.6) and (1.9), we obtain the equation of the Σ surface, i.e.,

$$f(q, E, c) = |E|t - p\pi/k \tag{1.11}$$

where p is an integer. It follows that the family of wave surfaces is of the form:

$$f(q, E, c) = \kappa \tag{1.12}$$

where κ is a variable parameter.

The curves which results from the time independent Hamilton−Jacobi equation, are denoted by C. Since the velocity of a point of the C curve, which results from the time independent Hamilton−Jacobi equation, is $\bar{v} = (1/m)\nabla f(q, E, c)$ (Landau and Lifschitz, 2000), it follows that the

C curve is normal to the Σ surface. The Σ surfaces and the C curves of the system are situated inside the CA domain.

The action of the system is

$$S = S_0(q, E, c) - Et = f(q, E, c) + K - Et \qquad (1.13)$$

where S_0 is the reduced action and K is a constant, whose expression will be calculated later (see Eq. (1.22)).

In virtue of the theory from Landau and Lifschitz (2000, p. 149)], the C curves depend on $6N$ constants: $c = (c_1, c_2, \ldots, c_{3N-1})$, $c_{3N} = E$, $d = (d_1, d_2, \ldots, d_{3N})$, with

$$d_j = \frac{\partial S_0(q, E, c)}{\partial c_j} \quad \text{for} \quad j = 1, 2, \ldots, 3N - 1 \quad \text{and} \quad d_{3N} = -t_0 \quad (1.14)$$

where t_0 is the initial time.

Thus, we have obtained an accurate connection between the Hamilton—Jacobi (1.10) and the Schrödinger (1.3) equations, because the wave surfaces are solutions of the Hamilton—Jacobi equation. This connection results without any approximation. Similar connections, which have been derived through entirely different methods, are presented in literature (Courant and Lax, 1956; Luis, 2003), where it is shown that the discontinuities of the partial second derivatives of the wave function propagate following the trajectories determined by the Hamilton—Jacobi equation, written for the same system. On the other hand, according to the theory of differential equations (Courant and Hilbert, 1962; Smirnov, 1984; Zauderer, 1983), these discontinuities occur across characteristic surfaces of the wave equation, resulting the similarity between our approach and the approaches presented in Courant and Lax (1956) and Luis (2003).

There are similarities between our approach and the Synge's approach (Synge, 1954), who developed a geometrical theory of mechanical systems, starting from the basic equations of the Hamiltonian mechanics. According to this theory, the mechanical system is characterized by trajectories (named by Synge "rays") and associated normal surfaces (named "waves"). Synge assumed that these surfaces are de Broglie waves. The Bohr quantization relation was derived from the periodicity properties of the system. Although the analysis of Synge is made in the frame of classical mechanics, while our analysis is made in the frame of quantum mechanics, the conclusions of both approaches are the same. In fact, we proved that the supposition of Synge, that the surface associated to the classical motion is the wave surface of de Broglie wave, is rigorously exact.

1.4 PERIODIC MOTION OF THE WAVE SURFACES

1.4.1 Properties of the Wave Surfaces

We present now the properties of the Σ wave surfaces which are associated to the motion on a C curve. We present first some relations which result directly from the theory of the Hamilton–Jacobi equation (Landau and Lifschitz, 2000). We consider a point P moving on a C curve, and the associated Σ surface normal to C at P (Figure 1.1). The coordinates of the point P are functions $q = q(s)$ of the parameter s, the distance along the C curve. The s values are assigned to each point of the curve C, and by convention, the sense of increasing of s is the same as the sense of increasing of f (remembering that, in virtue of Eq. (1.11), f increases continuously).

Since f is determined up to an arbitrary constant, we choose the following initial condition:

$$f(q, E, c) = 0 \quad \text{for} \quad t = 0 \tag{1.15}$$

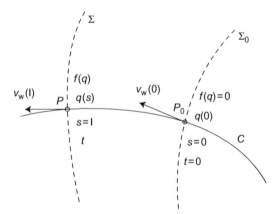

Figure 1.1 The motion of the point P together with the Σ surface along the C curve, between two points of the curve, corresponding, respectively, to $t = 0$, $q(0)$, $f = 0$ and t, $q(s)$, f, where s is the distance along the curve.

By Eqs. (1.10) and (1.11), the following relations are valid:

$$\frac{df}{ds} = |E|\frac{dt}{ds} \tag{1.16}$$

$$\frac{df}{ds} = \sqrt{\sum_j \left(\frac{\partial f}{\partial q_j}\right)^2} = \sqrt{2m(E - U)} \tag{1.17}$$

The velocity of the point P is equal to v_w, given by Eq. (1.5). This happens because from Eqs. (1.5), (1.16), and (1.17) (remembering that we consider the case corresponding to the plus sign in Eq. (1.5)), we have

$$\frac{ds}{dt} = v_w \tag{1.18}$$

Taking into account Eq. (1.15) and integrating Eq. (1.17) between P_0 and P, we find that at any time t before the surface crosses the initial position, the point P is situated on the Σ surface having the equation

$$f(q) = |E|t = \int_0^l \sqrt{2m(E - U)}ds \tag{1.19}$$

where l is the distance along the C curve, corresponding to the time t. The Σ surface, which is given by Eq. (1.19), is the wave surface associated with the motion on the C curve. In Section 1.7.2, we show examples of calculation of such surfaces in simple cases. We prove that these surfaces have the following properties.

Property 1.1

We fix a Σ surface, corresponding to constants c and E. Through a given point of this surface, denoted by P, and having the coordinates $q(P)$, passes only one C curve, whose constants are c, E, and d, where d is given by Eq. (1.14).

Proof

Through the point P passes only the curve whose constants d are $d_j = \partial S_0[q(P), E, c]/\partial c_j$, for $j = 1, 2, 3, \ldots, 3N - 1$ and $d_{3N} = -t_0$. The property follows.

Property 1.2

Two surfaces, Σ_1 and Σ_2, that correspond to the same values of the constants c and E, and have, respectively, the equations $f(q, E, c) = \kappa_1$ and $f(q, E, c) = \kappa_2$ are either non-intersecting (when $\kappa_1 \neq \kappa_2$ or identical (when $\kappa_1 = \kappa_2$). In other words, through a point of the CA domain passes at most one characteristic surface.

Proof

If Σ_1 and Σ_2 have a common point q, then $f(q, E, c) = \kappa_1 = \kappa_2$ and the property follows.

Property 1.3

If the moving Σ surface passes at the moment t_0 through a given surface Σ_0 and at a later time t_1 it intersects again the surface Σ_0, then at the time t_1 passes exactly through the surface Σ_0, moving in the same sense.

Proof

In virtue of Property 1.2, the Σ surface passes again exactly through its initial position Σ_0. We consider a point $P \in \Sigma \cap C$. By virtue of Eq. (1.11), we have $df = |E| dt$, resulting that the point P moves together with the surface only in the sense of increasing the values of the function f, on the C curve. Since the f function is single valued and its values are fixed on the CA domain, it follows that the sense of motion of the Σ surface is the same, anytime when it passes through the fixed surface Σ_0, and the property follows.

Property 1.4

The velocity v_w, corresponding to the plus sign in Eq. (1.5), is always positive, it never passes through zero and the point P moves always in the sense in which the distance s increases.

Proof

The Σ surfaces and the C curve are situated inside the CA domain, where $E > U$ and the kinetic energy is always positive. In virtue of Eq. (1.18), and taking into account that $v_w > 0$, the property follows.

The analysis of the case corresponding to the minus sign in Eq. (1.5) and plus in Eq. (1.9) shows that there is another wave surface that moves in the opposite direction.

Property 1.5

The motion of the Σ surface is periodic, and the C curve is closed.

Proof

Let P_0 be a fixed point on the C curve, corresponding to $f = 0$ and $t = 0$, and let P be a point moving along C curve together with the Σ surface (see Figure 1.1). In virtue of the above relations, at any time t before the surface crosses the initial position, the point P is situated on the Σ surface having Eq. (1.19).

If the Σ surface having Eq. (1.19) intersects again its initial position (having equation $f(q) = 0$), then Property 1.3 implies that the Σ surface passes exactly through the initial position, moving in the same sense. On the other hand, in virtue of Property 1.1, the C curve passes again through the point P_0. Letting τ_w be the minimal time after which the Σ surface intersects itself, it follows that the motion of the Σ surface is periodical of period τ_w, and that the C trajectory is closed. We will show that this is the case, proving thus the claim of the proposition.

Indeed, assume by contradiction that the moving Σ surface never intersects its initial position. In virtue of Eq. (1.19), the Σ surface passes through distinct positions in the CA domain, corresponding to different values of the distance s, along the C curve. Since it moves only in the sense of increasing of distance s, and its velocity never passes through zero (Property 1.4), it follows that the Σ surface scans a volume of the CA domain, whose measure increases continuously with t. Indeed, in Section 1.7.2, we will show that the dimensions of the Σ surface have the same order of magnitude as the dimensions of the CA domain. Therefore, we can choose a value of the time t, so that the measure of this volume can be higher than any given value, contradicting the fact that the Σ surface is situated inside the CA domain, which is finite. This contradiction proves the claim, and therefore the proposition.

1.4.2 Consequences of the Periodicity of the Wave

In Section 1.4.1, we have shown that the Σ surface moves periodically, with period τ_w, and from Eq. (1.19) it follows that

$$0 \leq f(q) < f_M \text{ where } f_M = |E| \cdot \tau_w \tag{1.20}$$

resulting that the function f is bounded.

The equation of the Σ wave surface can now be written explicitly:

$$f(q) = |E|t - p|E|\tau_w \qquad (1.21)$$

for $p\tau_w \leq t < (p+1)\tau_w$, with $p = 0, 1, 2, \ldots$, which is seen to agree with Eq. (1.11).

The reduced action function along the C curve, denoted by S_0, is given by the following equation (Landau and Lifschitz, 2000):

$$S_0(q, E, c) = f(q, E, c) + pf_M \qquad (1.22)$$

where $p = 0$ for $0 \leq t < \tau_w$, $p = 1$ for $\tau_w \leq t < 2\tau_w$. By comparison of Eqs. (1.13) and (1.22), it follows that $K = pf_M$, where K is the constant from Eq. (1.13).

We see from Eq. (1.22) that S_0 increases continuously along the curve C. The variation of the function S_0 along the closed curve C, denoted by $\Delta_C S_0$, is given by:

$$\Delta_C S_0 = f_M \qquad (1.23)$$

By virtue of Eqs. (1.20) and (1.23), we obtain:

$$\tau_w = \Delta_C S_0 / |E| \qquad (1.24)$$

A point P, on the C curve, moving together with the Σ wave surface, has velocity τ_w. On the other hand, the velocity which results from the Hamilton–Jacobi equation, on the same curve, in the point P, is given by the equation $E = U + mv^2/2$, resulting

$$v = \pm\sqrt{2(E - U)/m} \qquad (1.25)$$

From Eqs. (1.5) and (1.25), it follows that

$$vv_w = |E|/m \qquad (1.26)$$

We prove now the relation between the period of the motion along the C curve, denoted by τ, which results from the Hamilton–Jacobi equation, and the period τ_w of motion of a point P of the Σ wave surface, on the same curve. We consider the relations which result in the classical analysis from the virial theorem (Landau and Lifschitz, 2000):

$$E = \tilde{T} + \tilde{U} = \frac{1}{2}\tilde{U} = -\tilde{T} \qquad (1.27)$$

where T is the kinetic energy and the tilde denotes the average over a period according to the following relation:

$$\tilde{T} = \frac{1}{\tau}\int_C \sum_j \frac{p_j^2}{2m}\,dt = \frac{1}{2\tau}\oint_C \sum_j p_j\,dq_j = \frac{1}{2\tau}\oint_C dS_0 = \frac{\Delta_C S_0}{2\tau} \qquad (1.28)$$

From Eqs. (1.27) and (1.28), we have $\Delta_C S_0 = 2\tau|E|$, while from Eq. (1.24), $\Delta_C S_0 = \tau_w|E|$, from where it follows that

$$\tau_w = 2\tau \qquad (1.29)$$

From this relation, we see that the period τ of the classical motion is equal to the time after which two wave surfaces moving in opposite directions meet again, after starting in the same point.

1.5 GENERALIZED BOHR QUANTIZATION RELATION FOR THE C CURVES

1.5.1 The Integral Relation of the Schrödinger Equation

The time independent Schrödinger and Hamilton—Jacobi equations, valid for the system in discussion are

$$-\frac{\hbar^2}{2m}\sum_j \frac{\partial^2 \Psi_0}{\partial q_j^2} + (U - E)\Psi_0 = 0 \qquad (1.30)$$

$$\frac{1}{2m}\sum_j \left(\frac{\partial S_0}{\partial q_j}\right)^2 + U - E = 0 \qquad (1.31)$$

With the substitution

$$\Psi_0 = A\,\exp(i\sigma/\hbar) \qquad (1.32)$$

where σ is a complex valued function of the electron coordinates and A is an arbitrary constant, we write the Schrödinger equation in the following equivalent form:

$$\frac{1}{2m}\sum_j \left(\frac{\partial \sigma}{\partial q_j}\right)^2 + U - E - \frac{i\hbar}{2m}\sum_j \frac{\partial^2 \sigma}{\partial q_j^2} = 0 \qquad (1.33)$$

We integrate Eqs. (1.31) and (1.33) along the closed curve C in the time domain, corresponding to the classical motion. Averaging over the period of the classical motion τ, we obtain:

$$E - \frac{1}{\tau} \int_C U \, dt = \frac{1}{2m\tau} \int_C \sum_j \left(\frac{\partial S_0}{\partial q_j} \right)^2 dt \qquad (1.34)$$

$$E - \frac{1}{\tau} \int_C U \, dt = \frac{1}{2m\tau} \int_C \sum_j \left(\frac{\partial \sigma}{\partial q_j} \right)^2 dt - \frac{i\hbar}{2m\tau} \int_C \sum_j \frac{\partial^2 \sigma}{\partial q_j^2} \, dt \qquad (1.35)$$

Because the potential energy U in a point of the space of coordinates and the energy E are the same in both the classical and the quantum equations, from Eqs. (1.34) and (1.35) we obtain the following equation in σ:

$$\int_C \sum_j \left(\frac{\partial \sigma}{\partial q_j} \right)^2 dt - i\hbar \int_C \sum_j \frac{\partial^2 \sigma}{\partial q_j^2} \, dt = \int_C \sum_j \left(\frac{\partial S_0}{\partial q_j} \right)^2 dt \qquad (1.36)$$

In Appendix A, we prove the following relation:

$$I_C = \int_C \sum_j \frac{\partial^2 S_0}{\partial q_j^2} \, dt = 0 \qquad (1.37)$$

An integral relation admits a specific supplementary solution (in addition to the solutions coming from the initial equation which has been integrated), called generalized or weak solution (see, e.g.,, Zauderer, 1983, chapter 6). By virtue of Eq. (1.37), it follows that the generalized solution of Eq. (1.36) is the function

$$\sigma_g = S_0 \qquad (1.38)$$

By virtue of Eqs. (1.32) and (1.38), the time independent wave function corresponding to the generalized solution is

$$\Psi_{0g} = A \, \exp(iS_0/\hbar) \qquad (1.39)$$

The corresponding time dependent wave function results from Eqs. (1.2) and (1.39), as follows:

$$\Psi_g = A \, \exp(iS/\hbar) \quad \text{where} \quad S = S_0 - Et \qquad (1.40)$$

Since S and S_0 are, respectively, the action and reduced action corresponding to the C curve, it follows that Ψ_g is the wave function associated to the classical motion on the C curve.

1.5.2 Generalized Bohr Quantization Relations

Since Ψ_{0g} is single valued, it must have the same value when we go along the curve C and arrive again in the initial point. This property leads to the generalized Bohr quantization relation:

$$\Delta_C S_0 = nh \tag{1.41}$$

where n is a positive integer.

The analysis can be simplified in the case of systems for which the separation of variables is possible. In this case, the function S_0 can be written

$$S_0 = \sum_a S_{0a} \quad \text{where} \quad S_{0a} = S_{0a}(x_a, y_a, z_a)\big| \tag{1.42}$$

If we integrate now Eqs. (1.31) and (1.33) not on the curve C (like in Eqs. (1.34) and (1.35)), but on the C_a curve (the curve corresponding to the electron a), we obtain the following quantization relation:

$$\Delta_{C_a} S_{0a} = n_a h \tag{1.43}$$

n_a being the principal quantum number associated to the motion of the electron a, where $n_a = 1, 2, \ldots$.

From Eqs. (1.41)–(1.43), we obtain:

$$n = \sum_a n_a \tag{1.44}$$

resulting that the minimum value of n is N, the total number of electrons.

1.6 THE STATIONARITY CONDITION AND THE DE BROGLIE RELATIONS FOR MULTIDIMENSIONAL SYSTEMS

1.6.1 Proof of the Stationarity Condition

We analyze the motion of the point of the Σ surface, denoted by P, that belongs to the closed C curve, in a period. We denote by t_i, t_f, Ψ_i, Ψ_f, q_i, q_f, respectively, the initial and final moments of time, wave functions and corresponding coordinates (Figure 1.2).

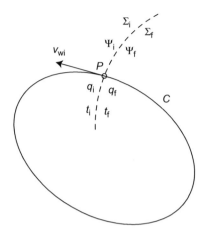

Figure 1.2 Illustration of the periodicity and stationarity of the wave (schematic).

By virtue of Eqs. (1.24) and (1.41), we have

$$|E|\tau_{\rm w} = nh \qquad (1.45)$$

From the periodicity of motion of the wave surface, it follows that $q_i \equiv q_f$ and $t_f = t_i + \tau_{\rm w}$. On the other hand, from Eq. (1.2), we have $\Psi_i = \Psi_0(q_i)\exp(-Et_i/\hbar)$ and $\Psi_f = \Psi_0(q_f)\exp(-Et_f/\hbar)$. Taking into account, these relations and Eq. (1.45), we obtain:

$$\Psi_i = \Psi_f \qquad (1.46)$$

This is the condition of the stationarity of the wave. It shows that, when the point P moves on the C curve and arrives again in the initial point, then the value of the wave function in that point comes back to the initial value.

1.6.2 Proof of the de Broglie Relations
Since Σ is the wave surface of the wave described by the Schrödinger equation, it follows that Σ is the *de Broglie wave surface* of the system. We prove now that this wave satisfies the generalized de Broglie relations in the space of the electron coordinates.

1.6.2.1 The First de Broglie Relation
The wavelength of the de Broglie wave is

$$\lambda_{\rm w} = \tau_{\rm w} \cdot v_{\rm w} \qquad (1.47)$$

From Eqs. (1.26) and (1.45), we have, respectively, $v_w = |E|/(mv)$ and $\tau_w = nh/|E|$ and, taking into account Eq. (1.47), we obtain the first de Broglie relation:

$$\lambda_w = n\frac{h}{mv} \tag{1.48}$$

which can be written:

$$\overline{p} = n\hbar\overline{k}_w \tag{1.49}$$

where $\overline{p} = m\overline{v}$ and $\overline{k}_w$ is the wave vector of the de Broglie wave, which is given by relation $\overline{k}_w = (2\pi/\lambda_w)(\overline{v}_w/v_w)$.

1.6.2.2 The Second de Broglie Relation

From Eq. (1.45), we obtain $|E| = nh/\tau_w$, from where it results

$$|E| = n\hbar\omega_w \tag{1.50}$$

where ω_w is the angular frequency of the de Broglie wave.

We recall that the theory of de Broglie waves (de Broglie, 1924, 1966; Wichmann, 1971) associates a wave to a moving particle. This wave is characterized by a wave vector $\overline{k}$ and an angular frequency ω, which are related to the momentum and energy of the particle, respectively, by the de Broglie relations, $\overline{p} = \hbar\overline{k}$ and $|E| = \hbar\omega$.

In this work, we showed that the de Broglie relations can be generalized for multidimensional systems, in the space of the electron coordinates. We proved that there is a de Broglie wave associated to the motion on the C curve, which results from the Hamilton—Jacobi equation. This wave is characterized by the wave vector $\overline{k}_w$ and the angular frequency ω_w. These quantities are related to the energy E and momentum $\overline{p}$, of the moving point on the C curve, by the de Broglie relations, which are Eqs. (1.49) and (1.50).

We note that the motion on the C curve is associated also with the wave function Ψ_g, which results from the integral relation of the Schrödinger equation on the C curve (see Eq. (1.40)).

In applications, we calculate the constants of motion resulting from the curves associated with each electron, which are obtained

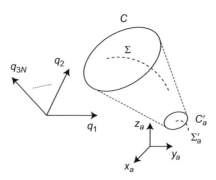

Figure 1.3 Projection of the Σ surface and C curve from the space R^{3N} on the space of the coordinates of the electron a (schematic).

by the projection of the curve C from the R^{3N} space of coordinates to the three-dimensional spaces of the electrons. For example, the curve of electron a is obtained from the projection of the curve C to the space of coordinates x_a, y_a, z_a, as shown in Figure 1.3. The projection is also a closed curve, denoted by C'_a, where $a = 1, 2, \ldots, N$. When the separation of variables is possible, we have $C'_a \equiv C_a$, where C_a is defined in Section 1.5.2.

The above properties imply that the problem of calculating the energetic values of atomic and molecular systems reduces to the calculation of the C'_a or C_a, or, in other words, to the calculation of the systems' periodic solutions. In Popa (Popa, 2008a, 2009a, 2011a), we have shown that the problem can be solved exactly, or with good accuracy, by a central field method, which makes possible the separation of the variables, and, consequently, the calculation of the C_a curves. In Popa (Popa, 2008a, 2009a, 2011a), the energy is calculated with the aid of Eq. (1.43), which is applied to the C_a curves. It follows that our central field method is, in fact, a Bohr-type method for multidimensional stationary systems.

1.7 PROPERTIES OF THE CENTRAL FIELD SEMICLASSICAL METHOD

1.7.1 The Constants of Motion of the System

In Section 1.2, we have shown that the constants of motion of the system described by the Schrödinger equation are the eigenvalues of the total energy E and the eigenvalues of the constants $\alpha = (\alpha_1, \alpha_2, \ldots, \alpha_S)$.

These eigenvalues have discrete values which depend on the quantum numbers of the system, according to

$$\alpha_j = \alpha_j(n_j) \tag{1.51}$$

where n_j is the quantum number which corresponds to the constant of motion having eigenvalue α_j.

For the approximation of the atomic orbitals resulting from the central field method, the number of quantum numbers is equal to the number of constants of motion, and to the number of coordinates of the system, resulting that $S = 3N - 1$. We briefly recall the theory leading to these facts.

In the case of a hydrogenoid system (for $N = 1$), there are three constants of motion: the energy, orbital momentum and its projection on the z-axis. The corresponding quantum numbers are, respectively, the principal, azimuthal, and magnetic orbital. In the expression of the wave function, the quantum numbers enter explicitly, while the constants of motion enter implicitly, because they are functions of the three quantum numbers.

For a general atomic system with N electrons, the wave function Ψ is function of N hydrogenoid wave functions, using the approximation of the atomic orbitals (Hartree, 1957; Slater, 1960; Coulson, 1961). Since each hydrogenoid wave function depends (explicitly) on three quantum numbers, and (implicitly) on three constants of motion, it follows that Ψ depends on $3N$ quantum numbers, corresponding to $3N$ constants of motion.

The case of a molecule with N' nuclei and N electrons is similar. Indeed, the linear combination of atomic orbitals (LCAO) approximation (Slater, 1963) shows that the wave function of the molecule is a function of N atomic wave functions, each of them corresponding to three constants of motion, and the same conclusion follows.

On the other hand, in Section 1.2 we have shown that the constants E and $c = (c_1, c_2, \ldots, c_{3N-1})$ which enter in the expression of the characteristic function and in the expression of the f function, are equal to the eigenvalues of the constants of motion of the system, resulting

$$c_j = \alpha_j \quad \text{for} \quad 1 \leq j \leq 3N - 1 \quad \text{and} \quad c_{3N} = E \tag{1.52}$$

1.7.2 Wave Surfaces for Hydrogenoid Systems

By the approximation of the central field method, the states of the electrons, and the C_a curves associated to these states are separated (Popa, 2008a, 2009a, 2011a). It follows that, with good approximation, the projections of the Σ surface and of the C curve on the space of coordinates of the electron a, are, respectively, the Σ_a surface and the C_a curve (see Figure 1.3). But the properties of the Σ_a surface and of the C_a curve can be exactly determined, since they are the same as for hydrogenoid systems. This is why it is useful to calculate the distributions of the Σ_a surfaces, corresponding to the C_a curve in a hydrogenoid system.

For simplicity, we consider a two-dimensional hydrogenoid system. In this case, the C curve is an ellipse and the Σ surfaces are plane curves. We suppose that the nucleus is on the center of the xoy Cartesian system and the C curve is an ellipse, whose major axis coincides with the ox axis, and minor axis is parallel to oy axis (Figure 1.4A). The minimum and maximum distances from the nucleus are denoted, respectively, by r_m and r_M.

The variation of the function f, between two points of the C curve, having the coordinates (x_0, y_0) and (x_1, y_1), is

$$f(x_1, y_1) - f(x_0, y_0) = \int_{x_0,y_0}^{x_1,y_1} \overline{v}\, d\overline{s} \tag{1.53}$$

We suppose that, in a first phase, a point denoted by Q moves from the point $A(r_m, 0)$ to the point $B(-r_M, 0)$, on the upper part of the ellipse, in the sense of the increasing of the distance r from nucleus.

Since f is determined up to an arbitrary constant, we choose the following initial condition:

$$f = 0 \quad \text{for} \quad x_0 = r_m \quad \text{and} \quad y_0 = 0 \tag{1.54}$$

In a second phase, the point Q moves on the lower part of the ellipse, from the point $B(-r_M, 0)$ to the point $A(r_m, 0)$, in the sense of the decreasing of the r distance.

In Appendix B, we calculate the equation of the wave surfaces, corresponding to the upper part of the ellipse. This equation, written with the aid of Eqs. (B.18) and (B.19),

$$\underline{f_1(x, y)} = \underline{\kappa} \tag{1.55}$$

where $\underline{\kappa}$ is a variable parameter.

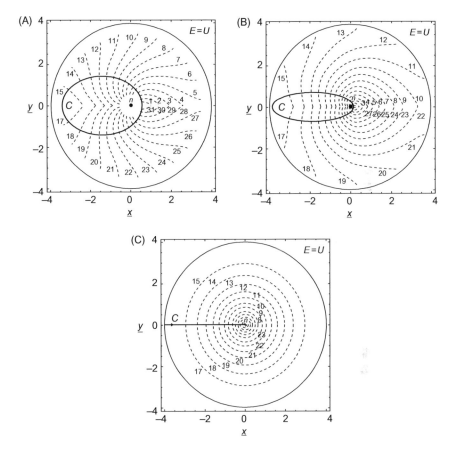

Figure 1.4 Σ (dashed lines), C curve and border of CA domain, for a bidimensional hydrogenoid system. The calculations are made for $Z = 1$ and $n = 2$, when $\underline{p}_\theta = \sqrt{2}$ (A), $\underline{p}_\theta = 0.5\sqrt{2}$ (B), and $\underline{p}_\theta = 0.001\sqrt{2}$ (C). The values of $\underline{\kappa}$ are given in the text.

The function f_1 depends on the parameters Z, n, e, r_m, r_M, and p_θ, where Z, n, e, and p_θ are, respectively, the order number of the nucleus, principal quantum number, eccentricity of the ellipse, and angular momentum. We use normalized quantities. The action and angular momentum are normalized to $\hbar$ and the distances to a_0. The normalized quantities are underlined, as follows: $\underline{f_1} = f_1/\hbar$, $\underline{p}_\theta = f_\theta/\hbar$, $\underline{r} = r/2a_0$, $\underline{r}_M = r_M/2a_0$, $\underline{r}_m = r_m/2a_0$, and so on. In Appendix B, we prove that

$$e = \sqrt{1 - \underline{p}_\theta^2/n^2}, \quad \underline{r}_m = \frac{n^2}{2Z}(1 - e)$$

and

$$\underline{r}_M = \frac{n^2}{2Z}(1+e)$$

In a similar mode, with the aid of relations (B.22) and (B.23), we obtain the equation of the wave surfaces, corresponding to the lower part of the ellipse, as follows:

$$\underline{f_2}(\underline{x}, \underline{y}) = \underline{\kappa} \tag{1.56}$$

In Appendix B, it is shown that the equation of the C curve is

$$\underline{b}^2(\underline{x} + \underline{a} - \underline{r}_m)^2 + \underline{a}^2\underline{y}^2 - \underline{a}^2\underline{b}^2 = 0 \tag{1.57}$$

where a and b are, respectively, the semi-major and semi-minor axes, whose normalized values are

$$\underline{a} = \frac{n^2}{2Z} \quad \text{and} \quad \underline{b} = \frac{n\underline{p}_\theta}{2Z} \tag{1.58}$$

The equation of the curve which represents the border of the CA domain is $E = U$, which can be written (see Appendix B):

$$\underline{x}^2 + \underline{y}^2 = \frac{n^4}{Z^2} \tag{1.59}$$

By virtue of Eq. (1.52), the values of p_θ are the eigenvalues of the constants of motion, which are given by the relation $p_\theta = \hbar\sqrt{l(l+1)}$ (Messiah, 1961), where l is the azimuthal quantum number.

In Figure 1.4A and B, we represent the C curve, the Σ wave surfaces (which are plane curves in this case) and the border of the CA domain, for 2p and 2s states. These states correspond, respectively, to $Z = 1$, $n = 2$, and $l = 1$ and to $Z = 1$, $n = 2$, and $l = 0$. Using the above relation, we find that the 2p state corresponds to $\underline{p}_\theta = \sqrt{2}$, while the 2s state corresponds to a negligible value of angular momentum, for which we consider that $\underline{p}_\theta = 0.001\sqrt{2}$. The C curve corresponding to the 2s state, shown in Figure 1.4C, is a pendular quasilinear ellipse. In order to see the evolution of the shape of the wave surfaces between the above two states, Figure 1.4B shows an intermediate situation, corresponding to $Z = 1$, $n = 2$, and $\underline{p}_\theta = 0.5\sqrt{2}$.

The Σ surfaces for the upper part of the C curve, which are numbered by $1, 2, \ldots, 15$ in Figure 1.4A, B, and C correspond, respectively, to $\underline{\kappa} = \pi/8$, $2\pi/8$, $\ldots$, $15\pi/8$. The Σ surfaces for the lower part of the C curve, which are numbered by $17, 18, \ldots, 31$, correspond, respectively, to $\underline{\kappa} = 17\pi/8$, $18\pi/8$, $\ldots$, $31\pi/8$. The calculations and the figures were obtained using Mathematica 7 program.

By the central field approximation, the states of the electrons and the C_a curves associated to these states, are separated. It follows that, with good approximation, the projections of the Σ surface and of the C curve on the space of coordinates of the electron a, are, respectively, the Σ_a surface and C_a curve (see Figure 1.3). But we have seen in this section that the dimensions of the Σ_a surface have the same order of magnitude as the dimension of the CA domain for the reduced system containing the electron a (see Figure 1.4A–C). It follows that the dimension of the Σ surface is of the same order of magnitude as the dimension of the CA domain.

CHAPTER 2

Connection Between Klein–Gordon and Relativistic Hamilton–Jacobi Equations for Systems Composed of Electromagnetic Fields and Particles

Abstract

We prove that the Klein-Gordon equation written for the system electron-very intense electromagnetic field is verified exactly by the wave function associated to the classical motion of the electron. We prove also that the components of the fields generated by interactions between very intense laser beams and electrons are periodic functions of only one variable, which is the phase of the incident field. These properties are proved without using any approximation, in the most general case in which the field is elliptically polarized and the initial phase of the incident field and the initial velocity of the electron are taken into consideration. These properties lead to exact models in the frame of the emergent field studying the interaction between very intense laser beams, on one side, and electron plasmas, relativistic electrons and atomic gases, on other side.

Keywords: Klein-Gordon equation; relativistic Hamilton-Jacobi equation; Liènard-Wiechert equation; electron-electromagnetic field system; periodicity property; Thomson scattering; high harmonics; very intense laser beams; electron plasmas; relativistic electron beam; atomic gas; hard radiations; angular distribution; spectral distribution; polarization effects

2.1 INITIAL HYPOTHESES

In this chapter, we review the theoretical results, referring to the connections between quantum and classical equations, in the case of the electrodynamic systems, presented in the papers by Popa (2004, 2007, 2011b, c, 2012). We recall that similar connections are presented, on a different way, in the papers by Motz (1962) and Motz and Selzer

(1964). We analyze a system composed of an electron interacting with a very intense electromagnetic elliptically polarized field. We consider the following initial hypotheses:

(h1) The interaction between the electron spin and the electromagnetic field is neglected and the behavior of the system is described by the Klein–Gordon equation (Messiah, 1961):

$$\left[c^2(-i\hbar\nabla + e\overline{A})^2 - \left(i\hbar\frac{\partial}{\partial t}\right)^2 + (mc^2)^2 \right]\Psi = 0 \qquad (2.1)$$

where $\overline{A}$ is the vector potential of the field, recalling that through-out the book, we denote by e the absolute value of the electron charge, the sign being written explicitly.

(h2) The electromagnetic field is elliptically polarized. In a Cartesian system of coordinates, the intensity of the electric field and of the magnetic induction vector, denoted respectively by $\overline{E}_L$ and $\overline{B}_L$, are polarized in the plane xy, while the wave vector, denoted by $\overline{k}_L$, is parallel to the oz axis. The expressions of the electric field and of the magnetic induction vector are as follows:

$$\overline{E}_L = E_{M1}\cos\eta\overline{i} + E_{M2}\sin\eta\overline{j} \qquad (2.2)$$

$$\overline{B}_L = -B_{M2}\sin\eta\overline{i} + B_{M1}\cos\eta\overline{j} \qquad (2.3)$$

with

$$\eta = \omega_L t - |\overline{k}_L|z + \eta_i \quad \text{and} \quad |\overline{k}_L|c = \omega_L \qquad (2.4)$$

where $\overline{i}$, $\overline{j}$, and $\overline{k}$ are versors of the ox, oy, and oz axes, E_{M1}, E_{M2}, B_{M1}, and B_{M2} are the amplitudes of the oscillations of the electromagnetic field components, ω_L is the angular frequency of the electromagnetic field, and η_i is an arbitrary initial phase. The following relations are also valid:

$$E_{M1} = cB_{M1}, \quad E_{M2} = cB_{M2} \quad \text{and} \quad c\overline{B}_L = \overline{k} \times \overline{E}_L \qquad (2.5)$$

Our theory applies also in the case of interactions between electrody-namic fields and electron plasmas, or electron beams. This is true, because if we consider a certain electron, denoted by e_a, the effect of the other electrons consists in the apparition of a supplementary potential, which is, with very good approximation, constant, supposing that the electron plasma and electron beam are homogeneous. It follows that both, quantum and classical equations of motion of e_a electron in the electromagnetic field are, with very good approximation, unchanged, compared with the case when the electron is isolated.

2.2 CONNECTION BETWEEN THE KLEIN—GORDON AND RELATIVISTIC HAMILTON—JACOBI EQUATIONS

We first obtained connections between Klein—Gordon and relativistic Hamilton—Jacobi equations and between Schrödinger and Hamilton—Jacobi equations, respectively, in Popa (2004, 2007), where we used the dipole approximation. We present now the approach from Popa (2011b), where we did not use any approximation.

With the substitution

$$\Psi = C \exp(i\sigma/\hbar) \tag{2.6}$$

where σ is a complex valued function of the electron coordinates and time and C is an arbitrary constant, the Klein—Gordon equation (2.1) becomes

$$c^2(\nabla\sigma + e\overline{A})^2 - \left(\frac{\partial\sigma}{\partial t}\right)^2 + (mc^2)^2 - i\hbar c^2\left[\nabla(\nabla\sigma + e\overline{A}) - \frac{\partial^2\sigma}{c^2\partial t^2}\right] = 0 \tag{2.7}$$

The relativistic Hamilton—Jacobi equation, written for the same system, is (Landau and Lifschitz, 1959)

$$c^2(\nabla S + e\overline{A})^2 - \left(\frac{\partial S}{\partial t}\right)^2 + (mc^2)^2 = 0 \tag{2.8}$$

where S is the classical action.

In the next section, we will show that for the systems considered we have

$$\nabla(\nabla S + e\overline{A}) - \frac{\partial^2 S}{c^2\partial t^2} = 0 \tag{2.9}$$

By Eqs. (2.7)–(2.9), we obtain the following property.

Property 2.1

Under hypotheses (h1) and (h2), the Klein–Gordon equation is verified by the wave function associated to the classical motion

$$\Psi = C \exp(iS/\hbar) \tag{2.10}$$

where S is the solution of the relativistic Hamilton–Jacobi equation, written for the same system.

The proof does not use any approximation.

A similar property is deduced in Motz (1962) and Motz and Selzer (1964), by a different method. The starting point there is the vanishing of the divergence of the energy-momentum four vector of the electron.

2.3 DEMONSTRATION OF THE RELATION (2.9)

In order to prove Eq. (2.9), we compute the expressions of the momentum of the electron and of $\partial S/\partial t$, starting with the classical equations of motion of the electron. Taking into account Eqs. (2.2) and (2.3), these equations are

$$m\frac{\mathrm{d}}{\mathrm{d}t}(\gamma v_x) = -eE_{M1}\cos\eta + ev_z B_{M1}\cos\eta \tag{2.11}$$

$$m\frac{\mathrm{d}}{\mathrm{d}t}(\gamma v_y) = -eE_{M2}\sin\eta + ev_z B_{M2}\sin\eta \tag{2.12}$$

$$m\frac{\mathrm{d}}{\mathrm{d}t}(\gamma v_z) = -ev_x B_{M1}\cos\eta - ev_y B_{M2}\sin\eta \tag{2.13}$$

where

$$\gamma = (1 - \beta_x^2 - \beta_y^2 - \beta_z^2)^{-\frac{1}{2}} \tag{2.14}$$

with $\beta_x = v_x/c$, $\beta_y = v_y/c$, and $\beta_z = v_z/c$.

We consider the following initial conditions in the most general case, when the components of the electron velocities have arbitrary values:

$$t = 0, \quad x = y = z = 0, \quad v_x = v_{xi}, \quad v_y = v_{yi}, \quad v_z = v_{zi}, \quad \text{and} \quad \eta = \eta_i \tag{2.15}$$

Taking into account Eq. (2.5), the equations of motion become

$$\frac{d}{dt}(\gamma\beta_x) = -a_1\omega_L(1 - \beta_z)\cos\eta \tag{2.16}$$

$$\frac{d}{dt}(\gamma\beta_y) = -a_2\omega_L(1 - \beta_z)\sin\eta \tag{2.17}$$

$$\frac{d}{dt}(\gamma\beta_z) = -\omega_L(a_1\beta_x \cos\eta + a_2\beta_y \sin\eta) \tag{2.18}$$

where

$$a_1 = \frac{eE_{M1}}{mc\omega_L} \quad \text{and} \quad a_2 = \frac{eE_{M2}}{mc\omega_L} \tag{2.19}$$

are relativistic parameters.

We calculate first β_x, β_y, and v_z, which are necessary in our analysis. We multiply Eqs. (2.16), (2.17), and (2.18), respectively, by β_x, β_y, and v_z. Taking into account that $\beta_x^2 + \beta_y^2 + \beta_z^2 = 1 - 1/\gamma^2$, their sum leads to

$$\frac{d\gamma}{dt} = -\omega_L(a_1\beta_x \cos\eta + a_2\beta_y \sin\eta) \tag{2.20}$$

From Eqs. (2.18) and (2.20), we obtain $d(\gamma\beta_z)/dt = d\gamma/dt$. We integrate this relation with respect to time between 0 and t, taking into account the initial conditions (2.15), and obtain $\gamma - \gamma_i = \gamma\beta_z - \gamma_i\beta_{zi}$. Using Eq. (2.4), we obtain

$$1 - \beta_z = \frac{1}{\omega_L} \cdot \frac{d\eta}{dt} = \frac{f_0}{\gamma} \tag{2.21}$$

with

$$f_0 = \gamma_i(1 - \beta_{zi}) \tag{2.22}$$

where $\beta_{xi} = v_{xi}/c$, $\beta_{yi} = v_{yi}/c$, and $\beta_{zi} = v_{zi}/c$ and $\gamma_i = 1/\left(1 - \beta_{xi}^2 - \beta_{yi}^2 - \beta_{zi}^2\right)^{1/2}$.

The integration of Eq. (2.16) with respect to time between 0 and t, taking into account Eq. (2.21) and the initial conditions (2.15) leads to the following relation:

$$\beta_x = \frac{f_1}{\gamma} \quad \text{with} \quad f_1 = -a_1(\sin\eta - \sin\eta_i) + \gamma_i\beta_{xi} \tag{2.23}$$

Similarly, integrating Eq. (2.17) and taking into account Eqs. (2.15) and (2.21), we obtain

$$\beta_y = \frac{f_2}{\gamma} \text{ with } f_2 = -a_2(\cos \eta_i - \cos \eta) + \gamma_i \beta_{yi} \qquad (2.24)$$

We substitute the expressions of β_x, β_y, and β_z, respectively, from Eq. (2.23), (2.24), and (2.21) into Eq. (2.14) and obtain the expression of γ:

$$\gamma = \frac{1}{2f_0}(1 + f_0^2 + f_1^2 + f_2^2) \qquad (2.25)$$

From Eq. (2.21), we have

$$\beta_z = \frac{f_3}{\gamma} \text{ with } f_3 = \gamma - f_0 \qquad (2.26)$$

According to the Lagrange–Hamilton formalism (Jackson, 1999; Landau and Lifschitz, 1959), the following relations are valid:

$$\overline{P} = \nabla S = \overline{p} - e\overline{A} \qquad (2.27)$$

$$H = -\frac{\partial S}{\partial t} = \gamma mc^2 \qquad (2.28)$$

where $\overline{P}$, $\overline{p}$, and H are, respectively, the generalized momentum of the electron, the momentum of the electron, and the Hamiltonian function of the electron.

By virtue of Eqs. (2.23)–(2.26), the quantities β_x, β_y, γ, and β_z are functions of only the phase η. On the other hand, from Eq. (2.4) it follows that η does not depend on x and y. With the aid of these properties, we obtain the expression of $\nabla \cdot \overline{p}$, as follows:

$$\nabla \cdot \overline{p} = \nabla(mc\gamma\beta_x\overline{i} + mc\gamma\beta_y\overline{j} + mc\gamma\beta_z\overline{k}) = mc\frac{\partial}{\partial z}(\gamma\beta_z)$$
$$= mc\frac{df_3}{d\eta} \cdot \frac{\partial\eta}{\partial z} = -mc|\overline{k}_L|\frac{d\gamma}{d\eta} \qquad (2.29)$$

Similarly, with the aid of Eqs. (2.4) and (2.28), obtain the expression of $\partial^2 S/(c^2\partial t^2)$:

$$\frac{\partial^2 S}{c^2\partial t^2} = -\frac{\partial H}{c^2\partial t} = -m\frac{\partial\gamma}{\partial t} = -m\frac{d\gamma}{d\eta} \cdot \frac{\partial\eta}{\partial t} = -m\omega_L\frac{d\gamma}{d\eta} \qquad (2.30)$$

Taking into account that $\omega_L = c|\bar{k}_L|$, we calculate the following expression, with the aid of Eqs. (2.27)—(2.30):

$$\nabla(\nabla S + e\bar{A}) - \frac{\partial^2 S}{c^2 \partial t^2} = \nabla \cdot \bar{p} - \frac{\partial^2 S}{c^2 \partial t^2} = 0 \qquad (2.31)$$

which proves Eq. (2.9). This finishes the proof of Property 2.1.

2.4 PERIODICITY PROPERTY OF THE SYSTEM ELECTRON—ELECTROMAGNETIC FIELD

From Property 2.1, it follows that a classical treatment of the system in discussion is justified. In Popa (2011b), we proved another property, which simplifies strongly the calculation of the angular and spectral distributions of the scattered radiation, emitted at the interaction between electrons and very intense laser beams. This property is as follows.

Property 2.2

The analytical expressions of electron velocity and acceleration, components of the electromagnetic field and the intensity of the beam generated by the relativistic interaction between electron and a plane electromagnetic field can be written in the form of composite periodic functions of only one variable, that is, the phase of the incident field.

This property is proved without using any approximation, in the most general case, in which the field is elliptically polarized, the initial phase of the incident field and the initial velocity of the electron are taken into consideration, and the direction in which the radiation is scattered is arbitrary. To prove this property, we use the relations from Section 2.3.

From Eqs. (2.20), (2.23), and (2.24), we have

$$\frac{d\gamma}{dt} = -\frac{\omega_L}{\gamma}(a_1 f_1 \cos \eta + a_2 f_2 \sin \eta) \qquad (2.32)$$

We calculate now the expressions of $\dot{\beta}_x$, $\dot{\beta}_y$, and $\dot{\beta}_z$. From Eq. (2.16), we obtain $\dot{\beta}_x$:

$$\dot{\beta}_x = -\frac{1}{\gamma}\left[\beta_x\frac{d\gamma}{dt} + a_1\omega_L(1 - \beta_z)\cos\eta\right] \tag{2.33}$$

By introducing in Eq. (2.33), the expressions of β_x, β_z, and $d\gamma/dt$, respectively from Eqs. (2.23), (2.21), and (2.32), we obtain

$$\dot{\beta}_x = \omega_L g_1 \text{ with } g_1 = -\frac{a_1 f_0}{\gamma^2}\cos\eta + \frac{f_1}{\gamma^3}(a_1 f_1\cos\eta + a_2 f_2\sin\eta) \tag{2.34}$$

Similarly, from Eqs. (2.17) and (2.18), we obtain $\dot{\beta}_y$ and $\dot{\beta}_z$:

$$\dot{\beta}_y = \omega_L g_2 \text{ with } g_2 = -\frac{a_2 f_0}{\gamma^2}\sin\eta + \frac{f_2}{\gamma^3}(a_1 f_1\cos\eta + a_2 f_2\sin\eta) \tag{2.35}$$

and

$$\dot{\beta}_z = \omega_L g_3 \text{ with } g_3 = -\frac{f_0}{\gamma^3}(a_1 f_1\cos\eta + a_2 f_2\sin\eta) \tag{2.36}$$

The analysis of relations (2.23)–(2.26) shows that f_1, f_2, γ, and f_3 are periodic functions of only one variable, that is η. Taking into account these relations, from Eqs. (2.34)–(2.36) it results that g_1, g_2, and g_3 are periodic functions of η. Finally, from Eqs. (2.23), (2.24), (2.26), (2.34)–(2.36) it follows, respectively, that β_x, β_y, β_z, $\dot{\beta}_x$, $\dot{\beta}_y$, and $\dot{\beta}_z$ are periodic functions of only one variable, η.

The periodicity of the scattered electromagnetic field results directly from the expression of the intensity of the electric field generated by the electron motion, which is given by the Liènard–Wiechert relation:

$$\overline{E} = \frac{-e}{4\pi\varepsilon_0 cR} \cdot \frac{\overline{n}\times[(\overline{n} - \overline{\beta})\times\dot{\overline{\beta}}]}{(1 - \overline{n}\cdot\overline{\beta})^3} \tag{2.37}$$

where R is the distance from the electron to the observation point (the detector) and $\overline{n}$ is the versor of the direction electron detector.

In virtue of the significance of the quantities entering in the Liènard—Wiechert equation, it results that the field $\overline{E}$ corresponds to the time $t + R/c$ and we have $\overline{E} = \overline{E}(\overline{r} + R\overline{n}, t + R/c)$, where $\overline{r}$ is the position vector of the electron with respect to a system having its origin at the point defined by Eq. (2.15). The relation $R \gg r$ is overwhelmingly fulfilled (Jackson, 1999). Using spherical coordinates for which θ is the azimuthal angle between the $\overline{n}$ and the $\overline{k}$ versors and ϕ is the polar angle in the plane xy, the components of the versor $\overline{n}$ can be written as follows:

$$n_x = \sin \theta \cos \phi, \quad n_y = \sin \theta \sin \phi \quad \text{and} \quad n_z = \cos \theta \qquad (2.38)$$

Introducing the components of $\overline{n}$, $\overline{\beta}$, and $\dot{\overline{\beta}}$ from Eqs. (2.23), (2.24), (2.26), (2.34)–(2.36), (2.38), we obtain the following expression of the intensity of the scattered electric field:

$$\overline{E} = \frac{K}{F_1^3}(h_1\overline{i} + h_2\overline{j} + h_2\overline{k}) \quad \text{with} \quad K = \frac{-e\omega_L}{4\pi\varepsilon_0 cR} \qquad (2.39)$$

where

$$h_1 = F_2\left(n_x - \frac{f_1}{\gamma}\right) - F_1 g_1 \qquad (2.40)$$

$$h_2 = F_2\left(n_y - \frac{f_2}{\gamma}\right) - F_1 g_2 \qquad (2.41)$$

$$h_3 = F_2\left(n_z - \frac{f_3}{\gamma}\right) - F_1 g_3 \qquad (2.42)$$

and

$$F_1 = 1 - n_x\frac{f_1}{\gamma} - n_y\frac{f_2}{\gamma} - n_z\frac{f_3}{\gamma} \qquad (2.43)$$

$$F_2 = n_x g_1 + n_y g_2 + n_z g_3 \qquad (2.44)$$

The magnetic induction vector of the scattered radiation is

$$\overline{B} = \overline{n} \times \frac{\overline{E}}{c} \qquad (2.45)$$

In virtue of Eq. (2.39), the intensity of the total scattered radiation can be written as follows:

$$I = c\varepsilon_0 \overline{E}^2 = c\varepsilon_0 K^2 \frac{1}{F_1^6}(h_1^2 + h_2^2 + h_3^2) \qquad (2.46)$$

Since $f_1, f_2, \gamma, f_3, g_1, g_2$, and g_3 are periodic functions of η, in virtue of Eqs. (2.39), (2.45), and (2.46), it follows that the components of the electromagnetic field and the intensity of the scattered beam are periodic functions of η. This finishes the proof of Property 2.2.

By virtue of Property 2.2, the components of the field $\overline{E}$ can be developed in Fourier series, and the expression of the field, normalized to K, becomes

$$
\begin{aligned}
\frac{\overline{E}}{K} = f_{1c0}\overline{i} &+ \left(\sum_{j=1}^{\infty} f_{1sj} \sin jn + \sum_{j=1}^{\infty} f_{1cj} \cos jn \right)\overline{i} \\
&+ f_{2c0}\overline{j} + \left(\sum_{j=1}^{\infty} f_{2sj} \sin jn + \sum_{j=1}^{\infty} f_{2cj} \cos jn \right)\overline{j} \qquad (2.47) \\
&+ f_{3c0}\overline{k} + \left(\sum_{j=1}^{\infty} f_{3sj} \sin jn + \sum_{j=1}^{\infty} f_{3cj} \cos jn \right)\overline{k}
\end{aligned}
$$

where

$$f_{\alpha c0} = \frac{1}{2\pi} \int_0^{2\pi} \frac{h_\alpha}{F_1^3}\, d\eta; \quad f_{\alpha sj} = \frac{1}{\pi} \int_0^{2\pi} \frac{h_\alpha}{F_1^3} \sin jn\, d\eta$$

$$f_{\alpha cj} = \frac{1}{\pi} \int_0^{2\pi} \frac{h_\alpha}{F_1^3} \cos jn\, d\eta \qquad (2.48)$$

with $\alpha = 1, 2, 3$.

The quantity

$$\overline{E}_j = \overline{E}_{js} + \overline{E}_{jc} \qquad (2.49)$$

where

$$\overline{E}_{js} = K(f_{1sj}\overline{i} + f_{2sj}\overline{j} + f_{3sj}\overline{k}) \sin jn \qquad (2.50)$$

$$\overline{E}_{jc} = K(f_{1cj}\overline{i} + f_{2cj}\overline{j} + f_{3cj}\overline{k})\cos jn \qquad (2.51)$$

is the intensity of the jth harmonic of the electric field. This harmonic is obtained by the addition of two plane fields having the same values of the angular frequency and wave vector magnitude, which are

$$\omega_j = j\omega_L \text{ and } \overline{k}_j = j\overline{k}_L \qquad (2.52)$$

The average intensity of the total scattered radiation, denoted by I_{av}, is given by the relation

$$I_{av} = \varepsilon_0 c \frac{1}{2\pi} \int_0^{2\pi} \overline{E}^2 \, d\eta \qquad (2.53)$$

With the aid of Eqs. (2.47), (2.53) and relations $\int_0^{2\pi} \cos m\eta \cos n\eta \, d\eta = \pi\delta_{mn}$, $\int_0^{2\pi} \sin m\eta \sin n\eta \, d\eta = \pi\delta_{mn}$, and $\int_0^{2\pi} \sin m\eta \cos n\eta \, d\eta = 0$, where m and n are integer numbers, the expression of the average intensity of the total scattered radiation, normalized to $\varepsilon_0 c K^2$, which is denoted by $\underline{I}_{av}$, becomes

$$\underline{I}_{av} = \frac{I_{av}}{\varepsilon_0 c K^2} = f_{1c0}^2 + f_{2c0}^2 + f_{3c0}^2 + \frac{1}{2}\sum_{j=1}^{\infty}(f_{1sj}^2 + f_{1cj}^2 + f_{2sj}^2 + f_{2cj}^2 + f_{3sj}^2 + f_{3cj}^2) \qquad (2.54)$$

The term $f_{1c0}^2 + f_{2c0}^2 + f_{3c0}^2$ corresponds to a constant component of the scattered field. The quantity

$$\underline{I}_j = \frac{1}{2}(f_{1sj}^2 + f_{1cj}^2 + f_{2sj}^2 + f_{2cj}^2 + f_{3sj}^2 + f_{3cj}^2) \qquad (2.55)$$

is the average intensity of the jth harmonic of the scattered field.

With the aid of Eq. (2.52), we obtain the following expressions of angular frequency, wavelength of the jth component of the scattered radiation and energy of the quanta of the scattered radiation, in an arbitrary direction:

$$\omega_j = j\omega_L, \quad \lambda_j = \lambda_L/j \text{ and } W_j = j\omega_L \hbar \qquad (2.56)$$

Property 2.2 is important because it makes possible to express physical quantities, like $\overline{E}$, $\underline{I}_{av}$, and $\underline{I}_j$, as composite functions of one variable that assume the form $f(\eta) = f(f_1(f_2(f_3(\ldots f_n(\eta)))))$. This strongly simplifies the calculation, because in this case it is not necessary to write explicitly $f(\eta)$, since its calculation reduces simply to successive calculations of functions $f_n, \ldots f_3$, f_2, f_1, and f, operations that can be performed numerically very fast and accurately. In our case, the initial data are a_1, a_2, β_{xi}, β_{yi}, β_{zi}, η_i, θ, ϕ, and j. The quantities $\overline{E}$, $\underline{I}_{av}$, and $\underline{I}_j$ are composite functions of f_1, f_2, γ, f_3, g_1, g_2, and g_3, which in their turn are periodic functions of η. It follows that the quantities $\overline{E}$, $\underline{I}_{av}$, and $\underline{I}_j$ can be calculated directly with basic mathematics software.

2.5 THE HEAD-ON INTERACTION BETWEEN VERY INTENSE ELLIPTICALLY POLARIZED LASER BEAMS AND RELATIVISTIC ELECTRON BEAMS, AS A SOURCE OF GENERATION OF VERY ENERGETIC RADIATIONS

2.5.1 Initial Data

From Property 2.1, it follows that the using of the relativistic Thomson scattering model is justified. The accuracy of this model is verified by numerous experimental data from literature, as it is shown also in our papers (Popa, 2008b, 2009b). A new version of the relativistic Thomson scattering has recently been developed, in which energetic radiations are generated by the head-on collision between very intense laser beams and relativistic electron beams (Anderson et al., 2004; Babzien, 2006; Kotaki et al., 2000; Pogorelsky et al., 2000; Sakai et al., 2003). Due to its potential importance, we analyze now this interaction.

We consider that an electromagnetic field described by Eqs. (2.2)–(2.5), which propagates in the oz direction, collides with a relativistic electron beam which propagates in the negative sense of the oz axis. It follows that the initial conditions, written in the laboratory reference system, denoted by $S(t, x, y, z)$, are as follows:

$$t = 0, \quad x = y = z = 0, \quad v_x = v_y = 0, \quad v_z = -|V_0|, \quad \text{and} \quad \eta = \eta_i \quad (2.57)$$

Since the phase of the electromagnetic field is a relativistic invariant, it is convenient to calculate the motion of the electron in the inertial system, denoted by $S'(t', x', y', z')$, in which the initial velocity of the electron is zero. It follows that the initial conditions in the S' system are as follows:

$$t' = 0, \quad x' = y' = z' = 0, \quad v'_{x'} = v'_{y'} = v'_{z'} = 0, \quad \eta' = \eta_i \tag{2.58}$$

where $v'_{x'}$, $v'_{y'}$, $v'_{z'}$, are the components of the electron velocity in the S' system.

2.5.2 Relativistic Motion of the Electron in the S' System

The Cartesian axes in the systems $S(t, x, y, z)$ and $S'(t', x', y', z')$ are parallel. In our case the S' system moves with velocity $-|\overline{V}_0|$ along the oz axis (Figure 2.1). Since our analysis is performed in the S' system, we have to calculate the parameters of the laser field, denoted by $\overline{E}'_L$, $\overline{B}'_L$, k'_L, and ω'_L, in the S' system.

The four-dimensional wave vectors are denoted, respectively, by $(\omega_L/c, k_{Lx}, k_{Ly}, k_{Lz})$ and $(\omega'_L/c, k'_{Lx'}, k'_{Ly'}, k'_{Lz'})$ in the systems S and S'.

In virtue of the Lorentz transformation, given by relations (11.22) of Jackson (1999), we have

$$\omega'_L = \gamma_0 \omega_L (1 + |\overline{\beta}_0|) \quad \text{and} \quad \omega'_L = c|\overline{k}'_L| \tag{2.59}$$

$$k'_{Lx'} = 0, \quad k'_{Ly'} = 0 \quad \text{and} \quad k'_{Lz'} = |\overline{k}'_L| = \gamma_0 |\overline{k}_L| (1 + |\overline{\beta}_0|) \tag{2.60}$$

where

$$\overline{\beta}_0 = -\frac{|\overline{V}_0|}{c} \overline{k} \quad \text{and} \quad \gamma_0 = (1 - \beta_0^2)^{-\frac{1}{2}} \tag{2.61}$$

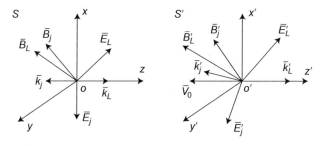

Figure 2.1 Components of the laser field and of the field generated by Thomson scattering S and S' systems.

The phase of the electromagnetic wave is a relativistic invariant (Jackson, 1999) and have

$$\eta = \omega_L t - \overline{k}_L \cdot \overline{r} + \eta_i = \omega'_L t' - \overline{k}'_L \cdot \overline{r}' + \eta_i = \eta' \qquad (2.62)$$

where $\overline{r}$ and $\overline{r}'$ are the position vectors of the electron in the systems S and S'. Since the relations between the space–time four-dimensional vectors in the systems S and S' are $ct' = \gamma_0(ct + |\overline{\beta}_0|z)$, $x' = x$, $y' = y$, and $z' = \gamma_0(z + |\overline{\beta}_0|ct)$, it is easy to see that these relations and Eqs. (2.59) and (2.60) verify Eq. (2.62).

We write equation (11.149) from Jackson (1999), which give the Lorentz transformation of the components of the electromagnetic field, in the International System, and using Eqs. (2.2), (2.3), and (2.62), we obtain the following expressions for the field components in the S' system:

$$\overline{E}'_L = \gamma_0(\overline{E}_L + \overline{\beta}_0 \times c\overline{B}_L) = \gamma_0(1 + |\overline{\beta}_0|)\overline{E}_L \qquad (2.63)$$

$$\overline{B}'_L = \gamma_0(\overline{B}_L - \overline{\beta}_0 \times \overline{E}_L/c) = \gamma_0(1 + |\overline{\beta}_0|)\overline{B}_L \qquad (2.64)$$

The equations of motion of the electron in the S' system are

$$m\frac{d}{dt'}(\gamma' v'_{x'}) = \gamma_0(1 + |\overline{\beta}_0|)(-eE_{M1}\cos\eta' + ev'_{z'}B_{M1}\cos\eta') \qquad (2.65)$$

$$m\frac{d}{dt'}(\gamma' v'_{y'}) = \gamma_0(1 + |\overline{\beta}_0|)(-eE_{M2}\sin\eta' + ev'_{z'}B_{M2}\sin\eta') \qquad (2.66)$$

$$m\frac{d}{dt'}(\gamma' v'_{z'}) = \gamma_0(1 + |\overline{\beta}_0|)(-ev'_{x'}B_{M1}\cos\eta' - ev'_{y'}B_{M2}\sin\eta') \qquad (2.67)$$

where $v'_{x'}$, $v'_{y'}$, and $v'_{z'}$ are the components of the electron velocity in the S' system, $\beta'_{x'} = v'_{x'}/c$, $\beta'_{y'} = v'_{y'}/c$, and $\beta'_{z'} = v'_{z'}/c$, while

$$\gamma' = (1 - \beta'^2_{x'} - \beta'^2_{y'} - \beta'^2_{z'})^{-\frac{1}{2}} \qquad (2.68)$$

In virtue of Eqs. (2.5) and (2.59), the system (2.65)–(2.67) becomes

$$\frac{d}{dt'}(\gamma'\beta'_{x'}) = -a'_1\omega'_L(1 - \beta'_{z'})\cos\eta' \qquad (2.69)$$

$$\frac{d}{dt'}(\gamma'\beta'_{y'}) = -a'_2\omega'_L(1 - \beta'_{z'})\sin\eta' \tag{2.70}$$

$$\frac{d}{dt'}(\gamma'\beta'_{z'}) = -\omega'_L(a'_1\beta'_{x'}\cos\eta' + a'_2\beta'_{y'}\sin\eta') \tag{2.71}$$

where

$$a'_1 = \frac{\gamma_0(1 + |\bar\beta_0|)eE_{M1}}{mc\omega'_L} = a_1 \text{ and } a'_2 = \frac{\gamma_0(1 + |\bar\beta_0|)eE_{M2}}{mc\omega'_L} = a_2 \tag{2.72}$$

The relativistic constants a_1 and a_2 are relativistic invariants and Eqs. (2.16)—(2.18) have the same form as Eqs. (2.69)—(2.71). The only difference is with respect to the initial conditions, which are

$$t' = 0, \quad x' = y' = z' = 0, \quad v'_{x'} = v'_{y'} = v'_{z'} = 0, \text{ and } \eta = \eta' = \eta_i \tag{2.73}$$

An identical procedure as that of Sections 2.3 and 2.4 leads to the following solution of the equation system (2.69)—(2.71):

$$\beta'_{x'} = \frac{f'_1}{\gamma'} \text{ with } f'_1 = -a_1(\sin\eta' - \sin\eta_i) \tag{2.74}$$

$$\beta'_{y'} = \frac{f'_2}{\gamma'} \text{ with } f'_2 = -a_2(\cos\eta_i - \cos\eta') \tag{2.75}$$

$$\gamma' = \frac{1}{2}(2 + f'^2_1 + f'^2_2) \tag{2.76}$$

$$\beta'_{z'} = \frac{f'_3}{\gamma'} \text{ with } f'_3 = \gamma' - 1 \tag{2.77}$$

$$\dot\beta'_{x'} = \omega'_L g'_1 \text{ with } g'_1 = -\frac{a_1}{\gamma'^2}\cos\eta' + \frac{f'_1}{\gamma'^3}(a_1f'_1\cos\eta' + a_2f'_2\sin\eta') \tag{2.78}$$

$$\dot\beta'_{y'} = \omega'_L g'_2 \text{ with } g'_2 = -\frac{a_1}{\gamma'^2}\sin\eta' + \frac{f'_2}{\gamma'^3}(a_1f'_1\cos\eta' + a_2f'_2\sin\eta') \tag{2.79}$$

$$\dot\beta'_{z'} = \omega'_L g'_3 \text{ with } g'_3 = -\frac{1}{\gamma'^3}(a_1f'_1\cos\eta' + a_2f'_2\sin\eta') \tag{2.80}$$

The analysis of the expressions of f'_1, f'_2, γ', f'_3, g'_1, g'_2, and g'_3 given, respectively, by Eqs. (2.74)—(2.80), shows that the quantities $\beta'_{x'}$, $\beta'_{y'}$, $\beta'_{z'}$, $\dot\beta'_{x'}$, $\dot\beta'_{y'}$, $\dot\beta'_{z'}$ are periodical functions of only one variable, that is the phase η'.

2.5.3 Spectral Components of the Electromagnetic Field
2.5.3.1 Spectral Components of the Electromagnetic Field in the S' System

The intensity of the electrical field generated by the motion of the electron, in the inertial S' system, is given by the Liènard–Wiechert relation:

$$\overline{E}' = \frac{-e}{4\pi\varepsilon_0 cR'} \cdot \frac{\overline{n}' \times [(\overline{n}' - \overline{\beta}') \times \dot{\overline{\beta}}']}{(1 - \overline{n}' \cdot \overline{\beta}')^3} \qquad (2.81)$$

where R' is the distance from the electron to the observation point (the detector) and $\overline{n}'$ is the versor of the direction electron detector. We calculate the right hand side of Eq. (2.81) at time t'. In virtue of the significance of the quantities entering in the Liènard–Wiechert equation, it results that the field $\overline{E}'$ corresponds to the time $t' + R'/c$ and we have $\overline{E}' = \overline{E}'(\overline{r}' + \overline{R}', t' + R'/c)$, where $\overline{R}' = R'\overline{n}'$. The inequality $R' \gg r'$ is strongly fulfilled (Jackson, 1999). We write the vector $\overline{n}'$ in a spherical coordinate system, so its components are written as follows:

$$n'_{x'} = \sin\theta' \cos\phi', \quad n'_{y'} = \sin\theta' \sin\phi', \quad n'_{z'} = \cos\theta' \qquad (2.82)$$

where θ' is the azimuthal angle between the $\overline{n}'$ and $\overline{k}$ versors and ϕ' is the polar angle in the plane $x'y'$.

Introducing the components of $\overline{\beta}'$ and $\dot{\overline{\beta}}'$, given respectively, by f'_1/γ', $f'_2/\gamma', f'_3/\gamma'$ and $\omega'_L g'_1, \omega'_L g'_2, \omega'_L g'_3$ in Eq. (2.81), we obtain the following expression of the intensity of the scattered electric field in the S':

$$\overline{E}' = \frac{K'}{F'^3_1}(h'_1 \overline{i} + h'_2 \overline{j} + h'_3 \overline{k}) \quad \text{with} \quad K' = \frac{-e\omega'_L}{4\pi\varepsilon_0 cR'} \qquad (2.83)$$

where

$$h'_1 = F'_2\left(n'_{x'} - \frac{f'_1}{\gamma'}\right) - F'_1 g'_1 \qquad (2.84)$$

$$h'_2 = F'_2\left(n'_{y'} - \frac{f'_2}{\gamma'}\right) - F'_1 g'_2 \qquad (2.85)$$

$$h'_3 = F'_2\left(n'_{z'} - \frac{f'_3}{\gamma}\right) - F'_1 g'_3 \qquad (2.86)$$

with

$$F_1' = 1 - n_{x'}' \frac{f_1'}{\gamma'} - n_{y'}' \frac{f_2'}{\gamma'} - n_{z'}' \frac{f_3'}{\gamma'} \tag{2.87}$$

$$F_2' = n_{x'}' g_1' + n_{y'}' g_2' + n_{z'}' g_3' \tag{2.88}$$

Since f_1', f_2', γ', f_3', g_1', g_2', and g_3' are periodic functions of η', from Eqs. (2.83)–(2.88) it follows that the components of the field given by Eq. (2.81) are periodic composite functions of η', and they can be developed in Fourier series. The periodicity of $\overline{E}'$ implies the periodicity of the intensity of the scattered beam, following that Property 2.2 is valid also in the case of the head-on collision between laser and relativistic electron beams.

We expand the components of the electric field $\overline{E}'$ in Fourier series and find the expression of the intensity of harmonics j of the electric field, which is denoted by $\overline{E}_j'$:

$$\overline{E}_j' = \overline{E}_{js}' + \overline{E}_{jc}' \tag{2.89}$$

where

$$\overline{E}_{js}' = K'(f_{1sj}'\bar{i} + f_{2sj}'\bar{j} + f_{3sj}'\bar{k})\sin j\eta' \tag{2.90}$$

$$\overline{E}_{jc}' = K'(f_{1cj}'\bar{i} + f_{2cj}'\bar{j} + f_{3cj}'\bar{k})\cos j\eta' \tag{2.91}$$

$$f_{\alpha sj}' = \frac{1}{\pi} \int_0^{2\pi} \frac{h_\alpha'}{F_1'^3} \sin j\eta' \, d\eta' \text{ and } f_{\alpha cj}' = \frac{1}{\pi} \int_0^{2\pi} \frac{h_\alpha'}{F_1'^3} \cos j\eta' \, d\eta' \tag{2.92}$$

with $\alpha = 1, 2, 3$.

From Eq. (2.81), we have $\overline{E}' \cdot \overline{n}' = 0$. Using this relation and Eq. (2.83), we obtain

$$h_1' n_{x'}' + h_2' n_{y'}' + h_3' n_{z'}' = 0 \tag{2.93}$$

We multiply Eqs. (2.90) and (2.91) by $\overline{n}'$, taking into account Eqs. (2.92) and (2.93) and obtain:

$$\overline{E}_{js}' \cdot \overline{n}' = 0, \ \overline{E}_{jc}' \cdot \overline{n}' = 0 \text{ and } \overline{E}_j' \cdot \overline{n}' = 0 \tag{2.94}$$

Since the phase of the j harmonics of the electromagnetic field is $j\eta'$, it follows that, in virtue of Eq. (2.62), the angular frequency and the wave vector of the radiation scattered in the $\bar{n}'$ direction, corresponding to the j harmonics, are

$$\omega'_j = j\omega'_L, \quad \bar{k}'_j = j|\bar{k}'_L|\bar{n}' \quad \text{and} \quad \omega'_j = c|\bar{k}'_L| \qquad (2.95)$$

The corresponding magnetic induction vectors are

$$\bar{B}'_j = \bar{n}' \times \frac{\bar{E}'_j}{c}, \quad \bar{B}'_{js} = \bar{n}' \times \frac{\bar{E}'_{js}}{c} \quad \text{and} \quad \bar{B}'_{jc} = \bar{n}' \times \frac{\bar{E}'_{jc}}{c} \qquad (2.96)$$

When the incident electromagnetic field is linearly polarized, the following property is valid.

Property 2.3

The following relations are valid in the case of head-on collision between a linearly polarized electromagnetic beam and a relativistic electron beam:

$$f'_{\alpha sj} = 0 \text{ when } j = 1, 3, 5, \ldots \text{ and } f'_{\alpha cj} = 0 \text{ when } j = 2, 4, 6, \ldots \qquad (2.97)$$

Proof
Since in this case $a_1 \neq 0$ and $a_2 = 0$, from Eqs. (2.75) and (2.79) it results that $f'_2 = g'_2 = 0$, the functions f'_1 and f'_3 are of the form $f(\sin \eta' - \sin \eta_i)$, and the functions g'_1 and g'_3 are of the form $f(\sin \eta' - \sin \eta_i)\cos \eta'$. From Eqs. (2.84)–(2.88), it follows that the functions h'_α/F'^3_1 are of the form $F(\eta') = f(\sin \eta' - \sin \eta_i)\cos \eta'$. Since the function F has the symmetry $F(\eta') = -F(\pi - \eta')$, it is easy to see that from Eq. (2.92) obtain Eq. (2.97). The proof is finished.

2.5.3.2 Spectral Components of the Electromagnetic Field in the S System

We use again the Lorentz transformations of the field given by equations (11.149) from Jackson (1999) to calculate the component $\bar{E}'_{js}$ in the laboratory system S. From Eq. (2.96), we obtain:

$$\begin{aligned}
\bar{E}_{js} &= \gamma_0(\bar{E}'_{js} - \bar{\beta}_0 \times c\bar{B}'_{js}) - \frac{\gamma_0^2}{\gamma_0 + 1}\bar{\beta}_0(\bar{\beta}_0 \cdot \bar{E}'_{js}) \\
&= \gamma_0(1 - |\bar{\beta}_0|\cos\theta')\bar{E}'_{js} + K'\gamma_0|\bar{\beta}_0|f'_{3sj}\sin j\eta'\left(\bar{n}' - \frac{\gamma_0|\bar{\beta}_0|}{\gamma_0 + 1}\bar{k}\right)
\end{aligned} \qquad (2.98)$$

Similarly, we have:

$$\overline{E}_{jc} = \gamma_0(1 - |\overline{\beta}_0| \cos \theta')\overline{E}'_{jc} + K'\gamma_0|\overline{\beta}_0|f'_{3cj} \cos j\eta' \left(\overline{n}' - \frac{\gamma_0|\overline{\beta}_0|}{\gamma_0 + 1}\overline{k}\right) \quad (2.99)$$

From Eqs. (2.82), (2.89)–(2.91), (2.98), and (2.99) and taking into account that the phase of the electromagnetic field is a relativistic invariant, namely $\eta = \eta'$, we obtain the expression for the intensity of the fundamental electrical field in the system S:

$$\overline{E}_j = \overline{E}_{js} + \overline{E}_{jc} \quad (2.100)$$

where

$$\overline{E}_{js} = K'(I_{1sj}\overline{i} + I_{2sj}\overline{j} + I_{3sj}\overline{k})\sin j\eta \quad (2.101)$$

$$\overline{E}_{jc} = K'(I_{1cj}\overline{i} + I_{2cj}\overline{j} + I_{3cj}\overline{k})\cos j\eta \quad (2.102)$$

with

$$I_{1sj} = \gamma_0(1 - |\overline{\beta}_0| \cos \theta')f'_{1sj} + \gamma_0|\overline{\beta}_0|\sin \theta' \cos \phi' f'_{3sj} \quad (2.103)$$

$$I_{1cj} = \gamma_0(1 - |\overline{\beta}_0| \cos \theta')f'_{1cj} + \gamma_0|\overline{\beta}_0| \sin \theta' \cos \phi' f'_{3cj} \quad (2.104)$$

$$I_{2sj} = \gamma_0(1 - |\overline{\beta}_0| \cos \theta')f'_{2sj} + \gamma_0|\overline{\beta}_0| \sin \theta' \sin \phi' f'_{3sj} \quad (2.105)$$

$$I_{2cj} = \gamma_0(1 - |\overline{\beta}_0| \cos \theta')f'_{2cj} + \gamma_0|\overline{\beta}_0| \sin \theta' \sin \phi' f'_{3cj} \quad (2.106)$$

$$I_{3sj} = f'_{3sj} \quad (2.107)$$

$$I_{3cj} = f'_{3cj} \quad (2.108)$$

In the particular case, when the incident field is linearly polarized, in virtue of Eqs. (2.97) and (2.103)–(2.108), we obtain

$$I_{1sj} = I_{2sj} = I_{3sj} = 0 \text{ and } \overline{E}_j = \overline{E}_{jc} \text{ when } j = 1, 3, 5, \dots \quad (2.109)$$

and

$$I_{1cj} = I_{2cj} = I_{3cj} = 0 \text{ and } \overline{E}_j = \overline{E}_{js} \text{ when } j = 2, 4, 6, \dots \quad (2.110)$$

2.5.4 Intensities of the Harmonic Radiation Beams in the S' and S Systems

The average value of the intensity of the jth component of the scattered radiation, normalized to $\varepsilon_0 c K'^2$, in the system S', is given by:

$$\underline{I}'_j = \frac{I'_j}{\varepsilon_0 c K'^2} = \frac{1}{2\pi} \int_0^{2\pi} \frac{\overline{E}'^2_j}{K'^2} d(j\eta') \tag{2.111}$$

Taking into account Eqs. (2.89)–(2.91), we obtain the following relation:

$$\underline{I}'_j = \frac{1}{2} ({f'_{1sj}}^2 + {f'_{1cj}}^2 + {f'_{2sj}}^2 + {f'_{2cj}}^2 + {f'_{3sj}}^2 + {f'_{3cj}}^2) \tag{2.112}$$

Similarly, the averaged value of the intensity of the fundamental component of the scattered radiation, normalized to $\varepsilon_0 c K'^2$, is given by:

$$\underline{I}_j = \frac{I_j}{\varepsilon_0 c K'^2} = \frac{1}{2\pi} \int_0^{2\pi} \frac{\overline{E}^2_j}{K'^2} d(j\eta) \tag{2.113}$$

Taking into account Eqs. (2.100)–(2.102) and (2.113), a simple calculation leads to the following relation:

$$\underline{I}_j = \frac{1}{2} (I^2_{1sj} + I^2_{1cj} + I^2_{2sj} + I^2_{2cj} + I^2_{3sj} + I^2_{3cj}) \tag{2.114}$$

We prove the following property which will be useful for the calculation of the intensities of the harmonic radiation beams in the S system.

Property 2.4

The following relations are valid:

$$\overline{E}^2_j = \gamma_0^2 (1 - |\overline{\beta}_0| \cos \theta')^2 \overline{E}'^2_j \tag{2.115}$$

$$\underline{I}_j = \gamma_0^2 (1 - |\overline{\beta}_0| \cos \theta')^2 \underline{I}'_j \tag{2.116}$$

where $\overline{E}'_j$ and $\overline{E}_j$ are the intensities of the electric field and $\underline{I}'_j$ and $\underline{I}_j$ are the beam intensities, corresponding to the j harmonics, in the S' and S systems.

Proof

In virtue of the Lorentz transformation of the electric field intensity from the S' to the S system, given by equation (11.149) of Jackson (1999), written in the International System, we have

$$\overline{E}_j^2 = \left[\gamma_0 (\overline{E}_j' - \overline{\beta}_0 \times c\overline{B}_j') - \frac{\gamma_0^2}{\gamma_0 + 1} \overline{\beta}_0 (\overline{\beta}_0 \cdot \overline{E}_j') \right]^2 \tag{2.117}$$

Since from Eqs. (2.94) and (2.96), we have $\overline{E}_j' \cdot \overline{n}' = 0$ and $c\overline{B}_j' = \overline{n}' \times \overline{E}_j'$, and $\overline{E}_j' \cdot [\overline{\beta}_0 \times (\overline{n}' \times \overline{E}_j')] = \overline{E}_j'^2 (\overline{\beta}_0 \cdot \overline{n}')$, Eq. (2.117) becomes

$$\overline{E}_j^2 = \gamma_0^2 \overline{E}_j'2 \left[1 + \overline{\beta}_0^2 \sin^2 \alpha_3 + \frac{\gamma_0^2 \overline{\beta}_0^4}{(\gamma_0 + 1)^2} \cos^2 \alpha_2 \right.$$
$$\left. + 2|\overline{\beta}_0| \cos \alpha_1 - \frac{2\gamma_0 \overline{\beta}_0^2}{\gamma_0 + 1} \cos^2 \alpha_2 \right] \tag{2.118}$$

where α_1 is the angle between $\overline{\beta}_0$ and $\overline{n}'$, α_2 is the angle between $\overline{\beta}_0$ and $\overline{E}_j'$, and α_3 is the angle between $\overline{\beta}_0$ and $\overline{n}' \times \overline{E}_j'$. It is easy to see that $\gamma_0^2 \overline{\beta}_0^2 / (\gamma_0 + 1)^2 - 2\gamma_0 / (\gamma_0 + 1) = -1$, and Eq. (2.118) can be written as follows:

$$\overline{E}_j^2 = \gamma_0^2 \overline{E}_j'^2 (1 + \overline{\beta}_0^2 \sin^2 \alpha_3 - \overline{\beta}_0^2 \cos^2 \alpha_2 + 2|\overline{\beta}_0| \cos \alpha_1) \tag{2.119}$$

The three vectors, $\overline{n}'$, $\overline{E}_j'$, and $\overline{n}' \times \overline{E}_j'$ form a right trihedral angle, therefore, $\cos^2 \alpha_1 + \cos^2 \alpha_2 + \cos^2 \alpha_3 - 1$, and we obtain

$$\overline{E}_j^2 = \gamma_0^2 (1 + |\overline{\beta}_0| \cos \alpha_1)^2 \overline{E}_j'^2 \tag{2.120}$$

Since $\alpha_1 = \pi - \theta'$, from Eq. (2.120), we obtain Eq. (2.115).

The intensity of the jth harmonics of the scattered radiation is $\varepsilon_0 c \overline{E}_j^2$ in the S system. Its averaged value, normalized to $\varepsilon_0 c K'^2$, results from Eqs. (2.113), (2.115), and (2.89–2.91), as follows:

$$I_j = \frac{I_j}{\varepsilon_0 c K'^2} = \gamma_0^2 (1 - |\overline{\beta}_0| \cos \theta')^2 \frac{1}{2\pi} \int_0^{2\pi} \frac{\overline{E}_j'^2}{K'^2} d(j\eta')$$
$$= \frac{1}{2} \gamma_0^2 (1 - |\overline{\beta}_0| \cos \theta')^2 \times (f_{1sj}'^2 + f_{1cj}'^2 + f_{2sj}'^2 + f_{2cj}'^2 + f_{3sj}'^2 + f_{3cj}'^2) \tag{2.121}$$

from where we obtain Eq. (2.116). The property is proved.

2.5.5 Relations Between Azimuthal and Polar Angles in the S and S' Systems and Energetic Relations

The four-dimensional wave vectors of the jth harmonics in the S and S' systems are denoted by $(\omega_j/c, k_{jx}, k_{jy}, k_{jz})$ and $(\omega_j'/c, k_{jx'}', k_{jy'}', k_{jz'}')$. With the aid of Eqs. (2.82) and (2.95), we have $k_{jx'}' = |\bar{k}_j'|\sin\theta'\cos\phi'$, $k_{jy'}' = |\bar{k}_j'|\sin\theta'\sin\phi'$, and $k_{jz'}' = |\bar{k}_j'|\cos\theta'$. In virtue of the Lorentz equations, the components of the wave vector in the S system are as follows:

$$\frac{\omega_j}{c} = \gamma_0\left(\frac{\omega_j'}{c} + \bar{\beta}_0 \cdot \bar{k}_j'\right) = \gamma_0\frac{\omega_j'}{c}(1 - |\bar{\beta}_0|\cos\theta') \tag{2.122}$$

$$k_{jz} = \gamma_0\left(k_{jz'}' - |\bar{\beta}_0|\frac{\omega_j'}{c}\right) = \gamma_0|\bar{k}_j'|(\cos\theta' - |\bar{\beta}_0|) \tag{2.123}$$

$$k_{jx} = |\bar{k}_j'|\sin\theta'\cos\phi' \quad \text{and} \quad k_{jy} = |\bar{k}_j'|\sin\theta'\sin\phi' \tag{2.124}$$

From Eqs. (2.122)–(2.124), we obtain

$$\frac{\omega_j}{c} = |\bar{k}_j| \tag{2.125}$$

We denote by $\bar{n}$ the versor of the direction in which the radiation is emitted in the S system. Its components are

$$n_x = \sin\theta\cos\phi, \quad n_y = \sin\theta\sin\phi \quad \text{and} \quad n_z = \cos\theta \tag{2.126}$$

From Eqs. (2.123) and (2.126), we have $k_{jz} = |\bar{k}_j|\cos\theta = \gamma_0|\bar{k}_j'|(\cos\theta' - |\bar{\beta}_0|)$. From Eqs. (2.95), (2.122), and (2.125), we obtain $|\bar{k}_j|/|\bar{k}_j'| = \gamma_0(1 - |\bar{\beta}_0|\cos\theta')$. From these relations, we deduce the relation between the angles θ and θ', as follows:

$$\cos\theta = \frac{\cos\theta' - |\bar{\beta}_0|}{1 - |\bar{\beta}_0|\cos\theta'} \quad \text{and} \quad \sin\theta = \frac{\sin\theta'}{\gamma_0(1 - |\bar{\beta}_0|\cos\theta')} \tag{2.127}$$

We note that these equations are identical to equation (5.6) from Landau and Lifschitz (1959), in spite of the fact that they are deduced in a completely different way.

In order to find the relation between ϕ and ϕ', we observe that, in virtue of Eqs. (2.124) and (2.126), we have $k_{jx} = |\bar{k}_j|\sin\theta\cos\phi = |\bar{k}_j'|\sin\theta'\cos\phi'$

and $k_{jy} = |\overline{k}_j| \sin \theta \sin \phi = |\overline{k}'_j| \sin \theta' \sin \phi'$. The ratio of these relations leads to

$$\phi = \phi' \tag{2.128}$$

It is easy to prove, that in virtue of Eqs. (2.98), (2.99), and (2.126)–(2.128), the following relations are valid:

$$\overline{E}_{js} \cdot \overline{n} = 0 \quad \text{and} \quad \overline{E}_{jc} \cdot \overline{n} = 0 \tag{2.129}$$

With the aid of Eqs. (2.59), (2.95), and (2.122), we obtain the following expressions of the angular frequency and of the wavelength of the jth component of the scattered radiation and energy of the quanta of the scattered radiation, in an arbitrary direction:

$$\omega_j = j\omega_L \gamma_0^2 (1 + |\overline{\beta}_0|)(1 - |\overline{\beta}_0| \cos \theta') \tag{2.130}$$

$$\lambda_j = \frac{\lambda_L}{j\gamma_0^2 (1 + |\overline{\beta}_0|)(1 - |\overline{\beta}_0| \cos \theta')} \tag{2.131}$$

$$W_j = \omega_j \hbar = j\omega_L \gamma_0^2 (1 + |\overline{\beta}_0|)(1 - |\overline{\beta}_0| \cos \theta')\hbar \tag{2.132}$$

The numerical calculation procedure is identical to that which has been described in Section 2.4, with the difference that in this case the initial data are a_1, a_2, η_i, θ', and ϕ'. Since $\overline{E}'_j$, $\overline{E}_j$, $\underline{I}'_j$, and $\underline{I}_j$ are composite functions of η, their calculation reduces to the successive calculations of the following functions: since f'_1, f'_2, γ', f'_3, g'_1, g'_2, g'_3, F'_1, F'_2, h'_1, h'_2, h'_3, f'_{1sj}, f'_{1cj}, f'_{2sj}, f'_{2cj}, f'_{3sj}, f'_{3cj}, I_{1sj}, I_{1cj}, I_{2sj}, I_{2cj}, I_{3sj}, and I_{3cj}. The calculation of $\underline{I}'_j$, $\underline{I}_j$, and W_j, given respectively, by Eqs. (2.112), (2.116), and (2.132), as functions of θ', is made with the aid of same Mathematica 7 program.

2.6 POLARIZATION EFFECTS IN THE INTERACTION, AT ARBITRARY ANGLE, BETWEEN VERY INTENSE LASER BEAMS AND RELATIVISTIC ELECTRON BEAMS

2.6.1 Initial Data

Another important particular system is composed of electron beam and electromagnetic beam, having a certain polarization. We consider the general case, when the angle between the two beams is arbitrary. For completeness, we consider both cases, of σ_L and π_L polarizations,

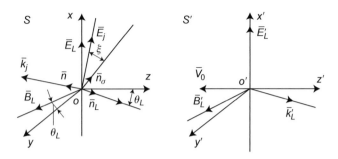

Figure 2.2 Components of the laser field in the S and S' systems. The incident field has the σ_L polarization. The intensities of electric fields, the magnetic induction vectors, and the wave vectors, for incident and scattered fields, are shown on figures.

when the linearly polarized electric field is perpendicular or parallel to the plane defined by the propagation directions of the laser beam and electron beam. These cases are described by the following relations:

a. The σ_L polarization, when the electric field of the incident beam is perpendicular on the plane defined by the wave vector of the incident beam and electron velocity denoted, respectively, by $\overline{k}_L$ and $\overline{V}_0$ (Figure 2.2A). The following relations, written in the laboratory system, are valid:

$$\overline{E}_L = E_M \cos \eta \overline{i}, \quad \overline{B}_L = B_M \cos \theta_L \cos \eta \overline{j} - B_M \sin \theta_L \cos \eta \overline{k} \quad (2.133)$$

$$\overline{k}_L = |\overline{k}_L|(\sin \theta_L \overline{j} + \cos \theta_L \overline{k}) \quad (2.134)$$

$$\overline{V}_0 = -|\overline{V}_0|\overline{k} \quad (2.135)$$

with

$$\eta = \omega_L t - \overline{k}_L \cdot \overline{r} + \eta_i, \quad |\overline{k}_L|c = \omega_L \quad (2.136)$$

$$E_M = cB_M, \quad c\overline{B}_L = (\overline{k}_L/|\overline{k}_L|) \times \overline{E}_L \quad (2.137)$$

b. The π_L polarization, when the electric field of the incident beam is parallel to the plane defined by the vectors $\overline{k}_L$ and $\overline{V}_0$ (Figure 2.3A). In this case, the components of the field become

$$\overline{E}_L = E_M \cos \theta_L \cos \eta \overline{i} - E_M \sin \theta_L \cos \eta \overline{k}, \quad \overline{B}_L = B_M \cos \eta \overline{j} \quad (2.138)$$

$$\overline{k}_L = |\overline{k}_L|(\sin \theta_L \overline{i} + \cos \theta_L \overline{k}) \quad (2.139)$$

while relations (2.135)–(2.137) remain valid.

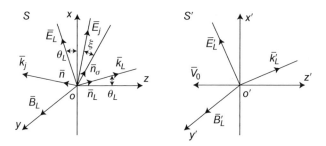

Figure 2.3 Components of the laser field in the S and S′ systems. The incident field has the π_L polarization. The intensities of electric fields, the magnetic induction vectors, and the wave vectors, for incident and scattered fields, are shown on figures.

The 180° and 90° geometries, in which the two beams collide, respectively, head-on and perpendicularly, are particular cases. These cases, correspond, respectively, to $\theta_L = 0$ and $\theta_L = \pm 90$, for both, σ_L and π_L polarizations.

The above relations are written in the laboratory reference system $S(t, x, y, z)$. Since the phase of the electromagnetic field is a relativistic invariant, it is convenient to calculate the motion of the electron, as in Section 2.5, in the inertial system $S'(t', x', y', z')$, in which the initial velocity of the electron is zero. It follows that the initial conditions, for both polarizations, in this system, are as follows:

$$t' = 0, \quad x' = y' = z' = 0, \quad v'_{x'} = v'_{y'} = v'_{z'} = 0, \quad \eta' = \eta_{\mathrm{i}} \qquad (2.140)$$

2.6.2 Relativistic Motion of the Electron in the S' System, When the Electromagnetic Field Is σ_L or π_L Polarized

2.6.2.1 The σ_L Polarization.

The Cartesian axes in the systems $S(t, x, y, z)$ and $S'(t', x', y', z')$ are parallel, and the S' system moves with velocity $-|\overline{V}_0|$ along the oz axis, as in Section 2.5 (see Figure 2.2). Since our analysis is performed in the S' system, we have to calculate the parameters of the laser field, denoted by $\overline{E}'_L$, $\overline{B}'_L$, $\overline{k}'_L$, and ω'_L, in the S' system.

The four-dimensional wave vectors are denoted, respectively, by $(\omega_L/c, k_{Lx}k_{Ly}, k_{Lz})$ and $(\omega'_L/c, k'_{Lx'}, k'_{Ly'}, k'_{Lz'})$ in the systems S and S'.

In virtue of the Lorentz transformation, given by relations (11.22) of Jackson (1999), we have

$$\omega'_L = \gamma_0 \omega_L (1 + |\bar{\beta}_0| \cos \theta_L) \text{ and } |\bar{k}'_L| = c\omega'_L \qquad (2.141)$$

$$k'_{Lx'} = 0, \ k'_{Ly'} = |\bar{k}_L| \sin \theta_L \text{ and } k'_{Lz'} = \gamma_0 |\bar{k}_L| (\cos \theta_L + |\bar{\beta}_0|) \qquad (2.142)$$

where $\bar{\beta}_0$ and γ_0 are given by Eq. (2.61).

The phase of the electromagnetic wave is a relativistic invariant (Jackson, 1999) and have

$$\eta = \omega_L t - \bar{k}_L \cdot \bar{r} + \eta_i = \omega'_L t' - \bar{k}'_L \cdot \bar{r}' + \eta_i = \eta' \qquad (2.143)$$

Since the relations between the space–time four-dimensional vectors in the systems S and S' are $ct' = \gamma_0(ct + |\bar{\beta}_0| z)$, $x' = x$, $y' = y$, and $z' = \gamma_0(z + |\bar{\beta}_0| ct)$, it is easy to see that these relations and Eqs. (2.141) and (2.142) verify Eq. (2.143).

We use again equation (11.149) from Jackson (1999), written in the International System, to calculate the Lorentz transformations of the field. Using Eq. (2.133), we obtain the following expressions for the field components in the S' system:

$$\bar{E}'_L = \gamma_0 (1 + |\bar{\beta}_0| \cos \theta_L) E_M \cos \eta' \bar{i} = E_M \frac{\omega'_L}{\omega_L} \cos \eta' \bar{i} \qquad (2.144)$$

$$\bar{B}'_L = [\gamma_0 (\cos \theta_L + |\bar{\beta}_0|) \bar{j} - \sin \theta_L \bar{k}] \frac{E_M}{c} \cos \eta' = (k'_{Lz'} \bar{j} - k'_{Ly'} \bar{k}) \frac{E_M}{c|\bar{k}_L|} \cos \eta' \qquad (2.145)$$

The equations of motion of the electron in the S' system are

$$m \frac{\mathrm{d}}{\mathrm{d}t'} (\gamma' v'_x) = -e E_M \frac{\omega'_L}{\omega_L} \cos \eta' + (k'_{Ly'} v'_{y'} + k'_{Lz'} v'_{z'}) \frac{e E_M}{c|\bar{k}_L|} \cos \eta' \qquad (2.146)$$

$$m \frac{\mathrm{d}}{\mathrm{d}t'} (\gamma' v'_{y'}) = -k'_{Ly'} v'_x \frac{e E_M}{c|\bar{k}_L|} \cos \eta' \qquad (2.147)$$

$$m \frac{\mathrm{d}}{\mathrm{d}t'} (\gamma' v'_z) = -k'_{Lz'} v'_x \frac{e E_M}{c|\bar{k}_L|} \cos \eta' \qquad (2.148)$$

Using Eqs. (2.136) and (1.141), the equations of motion can be written as follows:

$$\frac{d}{dt'}(\gamma'\beta'_{x'}) = -a\omega'_L\cos\eta' + a\omega'_L\left(\frac{k'_{Ly'}}{|\overline{k}'_L|}\beta'_{y'} + \frac{k'_{Lz'}}{|\overline{k}'_L|}\beta'_{z'}\right)\cos\eta' \quad (2.149)$$

$$\frac{d}{dt'}(\gamma'\beta'_{y'}) = -a\omega'_L\frac{k'_{Ly'}}{|\overline{k}'_L|}\beta'_{x'}\cos\eta' \quad (2.150)$$

$$\frac{d}{dt'}(\gamma'\beta'_{z'}) = -a\omega'_L\frac{k'_{Lz'}}{|\overline{k}'_L|}\beta'_{x'}\cos\eta' \quad (2.151)$$

where γ' is given by Eq. (2.68) and

$$a = \frac{eE_M}{mc\omega_L} \quad (2.152)$$

We multiply Eqs. (2.149), (2.150), and (2.151), respectively, by $\beta'_{x'}$, $\beta'_{y'}$, and $\beta'_{z'}$. Taking into account that $\beta'^2_{x'} + \beta'^2_{y'} + \beta'^2_{z'} = 1 - \gamma'^2$, their sum leads to:

$$\frac{d\gamma'}{dt'} = -a\omega'_L\beta'_{x'}\cos\eta' \quad (2.153)$$

From Eqs. (2.150) and (2.153), we obtain $d(\gamma'\beta'_{y'})/dt' = (k'_{Ly'}/|\overline{k}'_L|)d\gamma'/dt'$. Integrating this relation with respect to time between 0 and t', and taking into account the initial conditions (2.140), we have

$$\beta'_{y'} = \frac{k'_{Ly'}}{|\overline{k}'_L|}\left(1 - \frac{1}{\gamma'}\right) \quad (2.154)$$

Similarly, from Eqs. (2.151) and (2.153), we obtain

$$\beta'_{z'} = \frac{k'_{Lz'}}{|\overline{k}'_L|}\left(1 - \frac{1}{\gamma'}\right) \quad (2.155)$$

The differentiation of the phase η', given by Eq. (2.143), leads to:

$$\frac{d\eta'}{dt'} = \omega'_L - k'_{Ly'}c\beta'_{y'} - k'_{Lz'}c\beta'_{z'} = \omega'_L\left(1 - \frac{k'_{Ly'}}{|\overline{k}'_L|}\beta'_{y'} - \frac{k'_{Lz'}}{|\overline{k}'_L|}\beta'_{z'}\right) \quad (2.156)$$

and from (2.149) and (2.156), we obtain $d(\gamma'\beta'_{x'})/dt' = -a\cos\eta'\,d\eta'/dt'$. We integrate this relation with respect to time between 0 and t', and taking into the account the initial conditions (2.140), we obtain

$$\beta'_{x'} = \frac{f'_1}{\gamma'} \quad \text{with } f'_1 = -a(\sin\eta' - \sin\eta_i) \tag{2.157}$$

We substitute the expressions of $\beta'_{x'}$, $\beta'_{y'}$, and $\beta'_{z'}$, respectively, from Eqs. (2.157), (2.154), and (2.155) in $\beta'_{x'}{}^2 + \beta'_{y'}{}^2 + \beta'_{z'}{}^2 = 1 - \gamma'^2$ and obtain the expression of γ':

$$\gamma' = 1 + \frac{f'^2_1}{2} \tag{2.158}$$

From Eqs. (2.154) and (2.158) and, respectively, from Eqs. (2.155) and (2.158), we obtain the expressions of $\beta'_{y'}$ and $\beta'_{z'}$. With the aid of the relations (2.141) and (2.142), we have

$$\beta'_{y'} = \frac{f'_2}{\gamma'} \quad \text{with } f'_2 = \frac{\sin\theta_L}{\gamma_0(1 + |\bar{\beta}_0|\cos\theta_L)} \cdot \frac{f'^2_1}{2} \tag{2.159}$$

$$\beta'_{z'} = \frac{f'_3}{\gamma'} \quad \text{with } f'_3 = \frac{\cos\theta_L + |\bar{\beta}_0|}{1 + |\bar{\beta}_0|\cos\theta_L} \cdot \frac{f'^2_1}{2} \tag{2.160}$$

From Eq. (2.149), we obtain $\dot{\beta}'_{x'}$

$$\dot{\beta}'_{x'} = \frac{1}{\gamma'}\left[-\beta'_{x'}\frac{d\gamma'}{dt'} - a\omega'_L\cos\eta' + a\omega'_L\left(\frac{k'_{Ly'}}{|\bar{k}'_L|}\beta'_{y'} + \frac{k'_{Lz'}}{|\bar{k}'_L|}\beta'_{z'}\right)\cos\eta' \right] \tag{2.161}$$

From Eqs. (2.154) and (2.155), we obtain $k'_{Ly'}\beta'_{y'} + k'_{Lz'}\beta'_{z'} = |\bar{k}'_L|(1 - 1/\gamma')$. Introducing this expression in Eq. (2.161), together with the expressions of $d\gamma'/dt'$ and $\beta'_{x'}$, given respectively, by Eqs. (2.153) and (2.157), we obtain

$$\dot{\beta}'_{x'} = \omega'_L g'_1 \quad \text{with } g'_1 = \frac{a}{\gamma'^3}\left(\frac{f'^2_1}{2} - 1\right)\cos\eta' \tag{2.162}$$

Similarly, from Eqs. (2.150) and (2.151), we have

$$\dot{\beta}'_{y'} = \omega'_L g'_2 \quad \text{with } g'_2 = -\frac{\sin\theta_L}{\gamma_0(1 + |\bar{\beta}_0|\cos\theta_L)} \cdot \frac{af'_1}{\gamma'^3}\cos\eta' \tag{2.163}$$

$$\dot{\beta}'_{z'} = \omega'_L g'_3 \quad \text{with} \quad g'_3 = -\frac{\cos\theta_L + |\overline{\beta}_0|}{1 + |\overline{\beta}_0|\cos\theta_L} \cdot \frac{af'_1}{\gamma'^3}\cos\eta' \qquad (2.164)$$

The analysis of the expressions of f'_1, γ', f'_2, f'_3, g'_1, g'_2, and g'_3 given by Eqs. (2.157)–(2.160) and (2.162)–(2.164), reveals that the quantities $\beta'_{x'}$, $\beta'_{y'}$, $\beta'_{z'}$, $\dot{\beta}'_{x'}$, $\dot{\beta}'_{y'}$, $\dot{\beta}'_{z'}$ are periodical functions of only one variable, that is the phase η'.

2.6.2.2 The π_L Polarization
In this case, using Eqs. (2.136) and (2.139), we find that four-dimensional wave vector in the S system is $(\omega_L/c, |\overline{k}_L|\sin\theta_L, 0, |\overline{k}_L|\cos\theta_L)$. With the aid of the Lorentz transformations, we obtain the components of the four-dimensional vector in the S' system, denoted by $(\omega'_L/c, k'_{Lx'}, k'_{Ly'}, k'_{Lz'})$, as follows (see Figure 2.3):

$$\omega'_L = \gamma_0\omega_L(1 + |\overline{\beta}_0|\cos\theta_L) \quad \text{and} \quad |\overline{k}'_L| = c\omega'_L \qquad (2.165)$$

$$k'_{Lx'} = |\overline{k}_L|\sin\theta_L, \quad k'_{Ly'} = 0 \quad \text{and} \quad k'_{Lz'} = \gamma_0|\overline{k}_L|(\cos\theta_L + |\overline{\beta}_0|) \qquad (2.166)$$

where $\overline{\beta}_0$ and γ_0 are given by Eq. (2.61).

The relation (2.143) remains valid. Since the relations between the space–time four-dimensional vectors in the systems S' and S are the same as in the previous section, it is easy to see that these relations and Eqs. (2.165) and (2.166) verify Eq. (2.143).

With the aid of the Lorentz transformation of the fields, and using Eq. (2.138), we obtain the following expressions for the components of the electromagnetic field in the S' system:

$$\overline{E}'_L = \left[\gamma_0(\cos\theta_L + |\overline{\beta}_0|)\overline{i} - \sin\theta_L\overline{k}\right]E_M\cos\eta' = (k'_{Lz'}\overline{i} - k'_{Lx'}\overline{k})\frac{E_M}{|\overline{k}_L|}\cos\eta' \qquad (2.167)$$

$$\overline{B}'_L = \gamma_0\frac{E_M}{c}(1 + |\overline{\beta}_0|\cos\theta_L)\cos\eta'\overline{j} = \frac{E_M}{c}\cdot\frac{\omega'_L}{\omega_L}\cos\eta'\overline{j} \qquad (2.168)$$

With the aid of Eqs. (2.167) and (2.168), the equations of the motion of the electron can be written as follows:

$$m\frac{d}{dt'}(\gamma'v'_{x'}) = -eE_M\frac{\omega'_L}{\omega_L}\cdot\frac{k'_{Lz'}}{|\overline{k}'_L|}\cos\eta' + \frac{eE_M}{c}\cdot\frac{\omega'_L}{\omega_L}v'_z\cos\eta' \qquad (2.169)$$

$$m\frac{\mathrm{d}}{\mathrm{d}t'}(\gamma' v'_{z'}) = eE_M \frac{\omega'_L}{\omega_L} \cdot \frac{k'_{Lx'}}{|\overline{k}'_L|} \cos \eta' - \frac{eE_M}{c} \cdot \frac{\omega'_L}{\omega_L} v'_{x'} \cos \eta' \qquad (2.170)$$

where $v'_{y'} = 0$.

With the aid of Eq. (2.152), the equations of motion become

$$\frac{\mathrm{d}}{\mathrm{d}t'}(\gamma' \beta'_{x'}) = -a\omega'_L \frac{k'_{Lz'}}{|\overline{k}'_L|} \cos \eta' + a\omega'_L \beta'_{z'} \cos \eta' \qquad (2.171)$$

$$\frac{\mathrm{d}}{\mathrm{d}t'}(\gamma' \beta'_{z'}) = a\omega'_L \frac{k'_{Lx'}}{|\overline{k}'_L|} \cos \eta' - a\omega'_L \beta'_{x'} \cos \eta' \qquad (2.172)$$

with $\beta'_{y'} = 0$ and $\beta'^2_{x'} + \beta'^2_{z'} = 1 - \gamma'^2$.

We solve this system with the aid of the substitution

$$\beta'_{x'} = \beta_v \sin \xi + \beta_u \cos \xi \text{ and } \beta'_{z'} = \beta_v \cos \xi - \beta_u \sin \xi \qquad (2.173)$$

where

$$\sin \xi = \frac{k'_{Lx'}}{|\overline{k}'_L|} = \frac{\sin \theta_L}{\gamma_0(1 + |\overline{\beta}_0|\cos \theta_L)} \text{ and } \cos \xi = \frac{k'_{Lz'}}{|\overline{k}'_L|} = \frac{\cos \theta_L + |\overline{\beta}_0|}{1 + |\overline{\beta}_0|\cos \theta_L}$$
$$(2.174)$$

We substitute Eq. (2.173) in Eqs. (2.171) and (2.172), and, after substitution, multiply Eqs. (2.171) and (2.172), respectively, by $\cos \xi$ and $-\sin \xi$. The sum of these equations leads to Eq. (2.175). Similarly, by multiplication of the same equations by $\sin \xi$ and $\cos \xi$, we obtain Eq. (2.176).

$$\frac{\mathrm{d}}{\mathrm{d}t'}(\gamma' \beta_u) = -a\omega'_L(1 - \beta_v)\cos \eta' \qquad (2.175)$$

$$\frac{\mathrm{d}}{\mathrm{d}t'}(\gamma' \beta_v) = -a\omega'_L \beta_u \cos \eta' \qquad (2.176)$$

We multiply Eqs. (2.175) and (2.176), respectively, by β_u and β_v. Taking into account that $\beta_u^2 + \beta_v^2 = \beta'^2_{x'} + \beta'^2_{z'} = 1 - \gamma'^2$, their sum leads to:

$$\frac{\mathrm{d}\gamma'}{\mathrm{d}t'} = -a\omega'_L \beta_u \cos \eta' \qquad (2.177)$$

The differentiation of the phase η', given by Eq. (2.143), taking into account Eqs. (2.166), (2.173), and (2.174), leads to:

$$\frac{d\eta'}{dt'} = \omega'_L - k'_{Lx'}c\beta'_{x'} - k'_{Lz'}c\beta'_{z'} = \omega'_L(1 - \beta_v) \qquad (2.178)$$

From Eqs. (2.176) and (2.177), we have $d(\gamma'\beta_v)/dt' = d\gamma'/dt'$. We integrate this equation between 0 and t', taking into account the initial conditions (2.140) and Eq. (2.178), and have

$$\frac{1}{\gamma'} = 1 - \beta_v = \frac{1}{\omega'_L}\frac{d\eta'}{dt'} \qquad (2.179)$$

We introduce $1 - \beta_v$ from Eq. (2.179) in Eq. (2.175), and obtain $d(\gamma'\beta_u)/dt' = -a\cos\eta'\,d\eta'/dt'$. We integrate this equation between 0 and t', taking into the account of the initial conditions (2.140), and obtain

$$\beta_u = \frac{f}{\gamma'} \quad \text{with } f = -a(\sin\eta' - \sin\eta_i) \qquad (2.180)$$

From Eqs. (2.177) and (2.180), we obtain

$$\frac{d\gamma'}{dt'} = = -a\omega'_L\frac{f}{\gamma'}\cos\eta' \qquad (2.181)$$

From Eqs. (2.179) and (2.180), taking into account that $\beta_u^2 + \beta_v^2 = \beta_{x'}'^2 + \beta_{z'}'^2 = 1 - \gamma'^{-2}$, we have

$$\gamma' = 1 + \frac{f^2}{2} \qquad (2.182)$$

and from Eqs. (2.179) and (2.182), we obtain

$$\beta_v = \frac{f^2}{2\gamma'} \qquad (2.183)$$

Introducing Eqs. (2.180) and (2.183) in Eq. (2.173), we obtain the expressions of the normalized velocities, as follows:

$$\beta'_{x'} = \frac{f'_1}{\gamma'} \quad \text{with } f'_1 = \frac{\cos\theta_L + |\bar{\beta}_0|}{1 + |\bar{\beta}_0|\cos\theta_L}f + \frac{\sin\theta_L}{\gamma_0(1 + |\bar{\beta}_0|\cos\theta_L)}\cdot\frac{f^2}{2} \qquad (2.184)$$

$$\beta'_{z'} = \frac{f'_3}{\gamma'} \quad \text{with } f'_3 = -\frac{\sin\theta_L}{\gamma_0(1 + |\bar{\beta}_0|\cos\theta_L)}f + \frac{\cos\theta_L + |\bar{\beta}_0|}{1 + |\bar{\beta}_0|\cos\theta_L}\cdot\frac{f^2}{2}$$

$$(2.185)$$

From Eq. (2.171), we obtain $\dot{\beta}'_{x'}$:

$$\dot{\beta}'_{x'} = \frac{1}{\gamma'}\left(-\beta'_{x'}\frac{d\gamma'}{dt'} - a\omega'_L\frac{k'_{Lz'}}{|\bar{k}'_L|}\cos\eta' + a\omega'_L\beta'_{z'}\cos\eta'\right) \qquad (2.186)$$

Introducing the expressions of $d\gamma'/dt'$, $\beta'_{x'}$, and $\beta'_{z'}$, given respectively by Eqs. (2.181), (2.184), and (2.185), in Eq. (2.186), we obtain

$$\dot{\beta}'_{x'} = \omega'_L g'_1 \text{ with } g'_1 = \frac{a}{\gamma'}\left(\frac{ff'_1}{\gamma'^2} - \frac{f'_3}{\gamma'} + \frac{\cos\theta_L + |\bar{\beta}_0|}{1 + |\bar{\beta}_0|\cos\theta_L}\right)\cos\eta' \qquad (2.187)$$

$$\dot{\beta}'_{z'} = \omega'_L g'_3 \text{ with } g'_3 = \frac{a}{\gamma'}\left[\frac{ff'_3}{\gamma'^2} - \frac{f'_1}{\gamma'} + \frac{\sin\theta_L}{\gamma_0(1 + |\bar{\beta}_0|\cos\theta_L)}\right]\cos\eta' \qquad (2.188)$$

The analysis of the expressions of f, γ', f'_1, f'_3, g'_1, and g'_3 given, respectively, by Eqs. (2.180), (2.182), (2.184), (2.185), (2.187), and (2.188) reveals that the quantities $\beta'_{x'}$, $\beta'_{z'}$, $\dot{\beta}'_{x'}$, $\dot{\beta}'_{z'}$ are periodical functions of only one variable, that is the phase η'.

2.6.3 Periodicity Property and Energetic Relations

The relations which give the components of $\bar{\beta}'$ and $\dot{\bar{\beta}}'$ in the case of both, σ_L and π_L polarizations, namely $f'_1/\gamma', f'_2/\gamma', f'_3/\gamma', \omega'_L g'_1, \omega'_L g'_2$, and $\omega'_L g'_3$, are formally identical to those from Section 2.5. In addition, all these components are periodic functions of η'. In virtue of the Liènard–Wiechert equation, it follows that the components of the scattered electromagnetic field, $\bar{E}'$ and $\bar{B}'$, and the intensity of the total scattered radiation, I', are periodic functions of η', resulting that Property 2.2 is valid also in this case.

An analysis identical to that made in Section 2.5 shows that Eqs. (2.81)–(2.129) are valid also in the case of the interaction between laser and electron beams, at arbitrary angle, presented in this section.

But the energetic relations differ in this case. With the aid of Eqs. (2.95), (2.122), and (2.141), which are valid for both, σ_L and π_L polarizations, we obtain the following expressions of angular frequency, wavelength of the jth component of the scattered radiation and energy of the quanta of the scattered radiation, in an arbitrary direction:

$$\omega_j = j\omega_L\gamma_0^2(1 + |\overline{\beta}_0|\cos\theta_L)(1 - |\overline{\beta}_0|\cos\theta') \qquad (2.189)$$

$$\lambda_j = \frac{\lambda_L}{j\gamma_0^2(1 + |\overline{\beta}_0|\cos\theta_L)(1 - |\overline{\beta}_0|\cos\theta')} \qquad (2.190)$$

$$W_j = \omega_j\hbar = j\omega_L\gamma_0^2(1 + |\overline{\beta}_0|\cos\theta_L)(1 - |\overline{\beta}_0|\cos\theta')\hbar \qquad (2.191)$$

The numerical calculation procedure is identical to that which has been described at the end of Section 2.5.

2.6.4 Polarization of the Electromagnetic field of the Scattered Beam

In Popa (2012), we proved that by varying angle θ_L, the polarization of the scattered beam can be varied between the two limit configurations in which the electromagnetic field of the scattered beam is σ or π polarized with respect to the scattering plane. This result leads to the possibility to realize an adjustable photon source with both the energy and the polarization of the scattered radiations accurately controlled by the value of the θ_L angle.

The scattering plane is defined by the directions of the incident and scattered beams. The versors corresponding to these directions are $\overline{n}_L = \overline{k}_L/|\overline{k}_L|$ and $\overline{n}$. The versor normal to the scattering plane, which is the same as the versor of the electric field intensity, in the case of the σ polarization, denoted by $\overline{n}_\sigma$, is $\overline{n} \times \overline{n}_L$. Its expression is different for the σ_L and π_L polarizations of the incident beam, so we will analyze separately these cases.

2.6.4.1 The Case of the σ_L Polarization of the Incident Field

In virtue of relations (2.126) and (2.134), it follows that the versor normal to the scattering plane is

$$\overline{n}_\sigma = (\cos\theta_L n_y - \sin\theta_L n_z)\overline{i} - \cos\theta_L n_x\overline{j} + \sin\theta_L n_x\overline{k} \qquad (2.192)$$

In virtue of Eqs. (2.102), (2.109), and (2.192), it follows that the angle ξ_0 between the vector $\overline{E}_j$ and the direction of the σ polarization in the case of odd harmonics (when $\overline{E}_j = \overline{E}_{jc}$) is given by the following relation:

$$\cos \xi_o = \frac{\overline{E}_{jc}}{|\overline{E}_{jc}|} \cdot \overline{n}_\sigma = \frac{(\cos \theta_L n_y - \sin \theta_L n_z)I_{1cj} - \cos \theta_L n_x I_{2cj} + \sin \theta_L n_x I_{3cj}}{\sqrt{I_{1cj}^2 + I_{2cj}^2 + I_{3cj}^2}}$$

(2.193)

Similarly, the angle ξ_e between the vector $\overline{E}_j$ and the direction of the σ polarization, in the case of even harmonics (when $\overline{E}_j = \overline{E}_{js}$) is given by the relation:

$$\cos \xi_e = \frac{\overline{E}_{js}}{|\overline{E}_{js}|} \cdot \overline{n}_\sigma = \frac{(\cos \theta_L n_y - \sin \theta_L n_z)I_{1sj} - \cos \theta_L n_x I_{2sj} + \sin \theta_L n_x I_{3sj}}{\sqrt{I_{1sj}^2 + I_{2sj}^2 + I_{3sj}^2}}$$

(2.194)

2.6.4.2 The Case of the π_L Polarization of the Incident Field

From relations (2.126) and (2.139), we obtain the following expression of the versor normal to the scattering plane:

$$\overline{n}_\sigma = \cos \theta_L n_y \overline{i} + (\sin \theta_L n_z - \cos \theta_L n_x)\overline{j} - \sin \theta_L n_y \overline{k}$$ (2.195)

From Eqs. (2.102), (2.109), and (2.195), it follows that the angle between the vector $\overline{E}_j$ and the direction of the σ polarization, in the case of odd harmonics, is given by the relation:

$$\cos \xi_o = \frac{\overline{E}_{jc}}{|\overline{E}_{jc}|} \cdot \overline{n}_\sigma = \frac{\cos \theta_L n_y I_{1cj} + (\sin \theta_L n_z - \cos \theta_L n_x)I_{2cj} - \sin \theta_L n_y I_{3cj}}{\sqrt{I_{1cj}^2 + I_{2cj}^2 + I_{3cj}^2}}$$

(2.196)

Similarly, the angle between the vector $\overline{E}_j$ and the direction of the σ polarization, in the case of even harmonics, is given by the relation:

$$\cos \xi_e = \frac{\overline{E}_{js}}{|\overline{E}_{js}|} \cdot \overline{n}_\sigma = \frac{\cos \theta_L n_y I_{1sj} + (\sin \theta_L n_z - \cos \theta_L n_x)I_{2sj} - \sin \theta_L n_y I_{3sj}}{\sqrt{I_{1sj}^2 + I_{2sj}^2 + I_{3sj}^2}}$$

(2.197)

The relations (2.129) and (2.192)–(2.197) define completely the polarization of the scattered electromagnetic field $\overline{E}_j$. This happens because the vector $\overline{E}_j$ is situated in the plane perpendicular on $\overline{n}$, and it makes an angle ξ_o or ξ_e, respectively, for odd and even harmonics, with the direction of the σ polarization. If $\xi_o = 0$ or $\xi_e = 0$ the scattered electromagnetic field is σ polarized, while when $\xi_o = \pi/2$ or $\xi_e = \pi/2$ the scattered field is π polarized.

2.7 CLASSICAL APPROACH OF INTERACTIONS BETWEEN VERY INTENSE LASER BEAMS AND ATOMS

The development of very intense laser beams, corresponding to electrical fields of the order of an atomic unit, led to two complementary phenomena, which occur in interactions between laser beams and atoms. These phenomena, which are the above threshold ionization (ATI) and high harmonics generation (HHG), lead to the well-known sequential character of the electron motion, in which three phases can be distinguished. The first phase takes place in the atomic core domain, whose dimension is of the order of a few Bohr radii; it consists of the multiphoton absorbtion, after which the electron leaves the atom by tunneling. The second phase takes place in a very large ionization domain, whose dimension is of the order of thousands of Bohr radii, where the electrostatic potential of the nucleus is negligible and the electron is free. In the third phase, after oscillating in the electromagnetic field in the ionization domain, the electron returns in the atomic domain and transfers its energy to the field (high harmonic generation (HHG) phenomenon) or rescatters in the field of the nucleus.

A large number of semiclassical models of these phenomena lead to results in very good agreement with the experiments. We recall the classical model (Corkum, 1993), which leads to the most precise—to our knowledge—expression from literature for the cutoff of the high harmonic radiation. Property 2.1 justifies the accuracy of the Corkum's model and the relations that have been used for the demonstration of this property make possible a relativistic treatment of the motion of the electron in the ionization domain, in which the electrostatic interaction with nucleus is neglected, in the frame of the Corkum's model. This result will be used in applications to elaborate an accurate calculation method for the HHG spectrum.

Our analysis is made in the frame of the Corkum's model, which neglect both the length of the tunneling domain relative to the oscillation amplitude of the ionizing electron and the electron velocity at the beginning of the ionization domain. It follows that the initial conditions of the electron motion at the beginning of the ionization domain are

$$t = 0, \quad x = y = z = 0, \quad v_x = v_y = v_z = 0 \quad \text{and} \quad \eta = \eta_0 \qquad (2.198)$$

Since the electron comes back to the atomic core domain, and harmonics are emitted only when the laser field is linearly polarized, we limit the analysis to this case, when the intensity of the electric field and of the magnetic induction vector, are $\overline{E}_L = E_M \cos \eta \bar{i}$ and $\overline{B}_L = B_M \cos \eta \bar{j}$.

An identical solution of the system of equations of electron motion, as that presented in Section 2.3, leads to the following relations:

$$\beta_x = -\frac{a}{\gamma}(\sin \eta - \sin \eta_0), \quad \beta_z = \frac{a^2}{2\gamma}(\sin \eta - \sin \eta_0)^2 \qquad (2.199)$$

$$\beta_y = 0, \quad \text{and} \quad 1 - \beta_z = \frac{1}{\omega_L} \cdot \frac{d\eta}{dt} = \frac{1}{\gamma} \qquad (2.200)$$

where

$$a = \frac{eE_M}{mc\omega_L} \quad \text{and} \quad \gamma = 1 + \frac{a^2}{2}(\sin \eta - \sin \eta_0)^2 \qquad (2.201)$$

We integrate the first relation from Eq. (2.199) between 0 and t, which corresponds to the variation of η between η_0 and η, taking into account the second relation from Eq. (2.200) and the initial conditions (2.198) and obtain

$$x = \frac{ca}{\omega_L}\left[\cos \eta - \cos \eta_0 + \sin \eta_0(\eta - \eta_0)\right] \qquad (2.202)$$

In virtue of this relation, the electron returns to the atomic core, e.g., $x = 0$, when the phase of the field is equal to η_1, which satisfies

$$\cos \eta_1 - \cos \eta_0 + \sin \eta_0(\eta_1 - \eta_0) = 0 \qquad (2.203)$$

In virtue of Eq. (2.201), the kinetic energy of the electron in vicinity of nucleus, which corresponds to $\eta = \eta_1$, is (Landau and Lifschitz, 1959)

$$E_k = E - mc^2 = (\gamma - 1)mc^2 = \frac{mc^2 a^2}{2}(\sin \eta_1 - \sin \eta_0)^2 \qquad (2.204)$$

The ponderomotive energy is

$$U_p = \frac{e^2 E_M^2}{4m\omega_L^2} = \frac{1}{4}mc^2 a^2 \qquad (2.205)$$

Since the electromagnetic energy of the n order harmonics, which is emitted by electric dipole transition, in the frame of the Corkum model, is

$$E_{em} = E_k + I_p \qquad (2.206)$$

where I_p is the ionization energy, in virtue of Eqs. (2.204) and (2.205), it can be written as follows:

$$E_{em} = 2U_p(\sin \eta_1 - \sin \eta_0)^2 + I_p = n\hbar\omega_L \qquad (2.207)$$

By Eq. (2.203), we have $\eta_1 = \eta_1(\eta_0)$, and from Eqs. (2.203)–(2.207) it follows that, for a given value of η_0, there corresponds only one value for η_1, n, E_k, and E_{em}.

On the other hand, the kinetic momentum of the electron is

$$\overline{p} = mc\gamma\beta_x \overline{i} + mc\gamma\beta_z \overline{k} \qquad (2.208)$$

With the aid of Eqs. (2.199)–(2.201) and (2.205), the expression of $\overline{p}^2$, which will be used in applications, becomes

$$\overline{p}^2 = 4U_p m(\sin \eta_1 - \sin \eta_0)^2 \left[1 + \frac{a^2}{4}(\sin \eta_1 - \sin \eta_0)^2 \right] \qquad (2.209)$$

From Eq. (2.207) and relation $\eta_1 = \eta_1(\eta_0)$ it follows that a given n determines the value of η_0. In turn, as we have shown above, η_0 determines the values of η_1, E_k, and E_{em}. In Figure 2.4, we show the variations of $\eta_1/2\pi$ and $E_k/(3.17U_p)$ for the domain of η_0 for which the probability of the tunneling of the electron, denoted by W, is maximum. This probability is given (Corkum, 1993) by the relation $W = c_1(4\omega_s/\omega_t)^{2n^* - m_m - 1} \exp[-4\omega_s/(3\omega_t)]$, where $\omega_s = I_p/\hbar$, $\omega_t = eE_M \cos \eta_0 (2mI_p)^{-1/2}$, $n^* = (I_H/I_p)^{1/2}$, m_m is the magnetic quantum number, I_H is the ionization potential of hydrogen and c_1 is a constant which results from equation of W from Corkum (1993). Figure 2.4 also shows the variation of $10^4 W/c_1$ for the helium atom, when $I_p = 24.6$ eV and $m_m = 0$. Variations identical to those presented in Figure 2.4 are valid when η_0 varies around the values π, 2π, and so on.

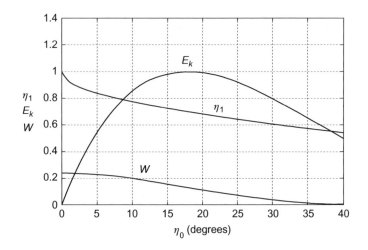

Figure 2.4 Normalized quantities $\eta_1/2\pi$, $E_k/(3.17U_p)$ and $10^4 W/c_1$, as functions of η_0. $\underline{W}$ is calculated for helium for $I_p = 24.6 eV$ and $m_m = 0$.

Since the cutoff energy of the spectrum, denoted by E_c, occurs when $\underline{E}_k$ is maximum, namely for $\eta_0 = 17°$ in Figure 2.4, we obtain the well-known Corkum formula:

$$E_c = 3.17U_p + I_p \qquad (2.210)$$

The above proof of Eq. (2.210) is different of the standard proof (Corkum, 1993), which is made in the non-relativistic case, when the dipole approximation is taken into account. We will use the relativistic relations proved in this section, to calculate the spectrum of high harmonics generated at interaction between very intense laser beams and atoms.

For the calculation of this spectrum we will use also the expression of W, normalized to its maximum value, which is

$$\underline{W} = \frac{1}{|\cos \eta_0|^{2n^* - m_m - 1}} \exp\left[-\left(\frac{1}{|\cos \eta_0|} - 1\right)\frac{4(2m)^{1/2}I_p^{3/2}}{3eE_M\hbar}\right] \qquad (2.211)$$

2.8 COMMON PROPERTIES OF THE SYSTEMS ANALYZED IN CHAPTERS 1 AND 2

Despite the fact that the stationary atomic and molecular (AM) systems and the nonstationary electrodynamic (ED) systems are very different, they have common properties.

The connection between quantum and classical equations results directly, without any supplementary postulate or approximation, from the properties of quantum equations, in both cases. So, the connection between Eqs. (1.3) and (1.10) in the case of AM systems, as well the connection between Eqs. (2.1) and (2.8) result directly.

In the same time, a classical solution of the type $A \exp(iS/\hbar)$ results, without any approximation, from the mathematical properties of both, the Schrödinger and Klein—Gordon equations (see Eqs. (1.40) and (2.10)). The classical solution in the case of the AM systems is associated with the geometric elements of the wave described by the Schrodinger equation, which result from the solution of the Hamilton—Jacobi equation, written for the same system. The classical solution in the case of the NE systems is associated to the classical electron trajectory which results from the relativistic Hamilton—Jacobi equation. There is a similitude between our result and the double solution theory of de Broglie, in spite of the fact that the two treatments are completely different. In the double solution theory (de Broglie, 1956), it is shown the existence of a second wave solution, which is a classical-type wave function, as in our analysis.

On the other hand, the existence of connections between quantum and classical equations is accompanied by the existence of periodicity properties in both cases, for AM and ED systems. This property has a practical importance, as we will show in the next chapters, because it leads to accurate models for the calculation of properties of AM and ED systems.

Proof of Equation (1.37)

We prove the relation:

$$I_C = \int_C \sum_j \frac{\partial^2 S_0}{\partial q_j^2} dt = 0 \tag{A.1}$$

By virtue of the relation (Landau and Lifschitz, 2000):

$$p_j = m v_j = m \frac{dq_j}{dt} = \frac{\partial f}{\partial q_j} = \frac{\partial S_0}{\partial q_j} \tag{A.2}$$

where p_j and v_j are the components of the momentum and velocity, we have

$$I_C = \int_C \sum_j \frac{\partial p_j}{\partial q_j} dt \tag{A.3}$$

On the other hand, the following relation is valid for all the functions which do not depend explicitly on time:

$$\frac{\partial}{\partial q_j} = \frac{d}{dq_j} - \frac{1}{v_j}\frac{\partial}{\partial t} = \frac{d}{dq_j} \tag{A.4}$$

This relation is valid for $S_0 = S_0(q)$ and for $p_j = \partial S_0/\partial q_j$. By virtue of Eqs. (A.2) and (A.4), the integral I_C becomes

$$I_C = \int_C \sum_j \frac{\partial p_j}{\partial q_j} dt = \int_C \sum_j \frac{dp_j}{v_j\, dt} dt = m \int_C \sum_j \frac{dp_j}{p_j} = m(\ln p_{jf} - \ln p_{ji}) = 0 \tag{A.5}$$

where p_{ji} and p_{jf} are, respectively, the initial and final values of the momentum components. These values are equal because the motion is periodic and the initial and final points are the same. The relation (A.1), or (1.37) from Section 1.4.1 is proved.

Relations for the Elliptic Motion. Equation of the Wave Surfaces for Hydrogenoid Systems

We consider the elliptic motion of an electron in a field of a nucleus with effective order number Z. In polar coordinates with the center at the nucleus, $x = r \cos \theta$ and $y = r \sin \theta$, the parametric equations of motion for electron can be written as follows (Landau and Lifschitz, 2000):

$$\frac{dr}{dt} = \pm \frac{1}{mr} \sqrt{\alpha r^2 + \beta r + \gamma} = \pm \frac{1}{r} \sqrt{\frac{2|E|}{m}} \sqrt{(r - r_m)(r_M - r)} \qquad (B.1)$$

$$\frac{d\theta}{dt} = \frac{p_\theta}{mr^2} \qquad (B.2)$$

where the constants α, β, γ, and K_1 are

$$\alpha = -2m|E|, \quad \beta = 2mK_1 Z, \quad \gamma = -p_\theta^2, \quad K_1 = \frac{e^2}{4\pi\varepsilon_0} \qquad (B.3)$$

while E, p_θ, r_m, and r_M are, respectively, the total energy, the angular momentum, and the minimum and maximum distances from the nucleus, which are given by the following relations:

$$E = -\frac{K_1 Z}{r} + \frac{mv^2}{2} \quad p_\theta = mr^2 \frac{d\theta}{dt} = \text{const.} \qquad (B.4)$$

$$r_m = a(1 - e) \quad \text{and} \quad r_M = a(1 + e) \qquad (B.5)$$

The ellipse is characterized by the quantities a, b, e, and u, which are, respectively, the semimajor axis, the semiminor axis, ellipticity, and the ellipse parameter. They are given by the following relations:

$$a = \frac{K_1 Z}{2|E|}, \quad b = \frac{p_\theta}{\sqrt{2m|E|}}, \quad e = \sqrt{1 - \frac{2|E|p_\theta^2}{mK_1^2 Z^2}}, \quad \text{and} \quad u = \frac{p_\theta^2}{mK_1 Z} \qquad (B.6)$$

The equation of the ellipse can be written as follows:

$$\frac{(x + a - r_m)^2}{a^2} + \frac{y^2}{b^2} - 1 = 0 \quad \text{or} \quad \frac{u}{r} = 1 - e \cos \theta \tag{B.7}$$

For quasipendular ellipses, for which $p_\theta \cong 0$, the following relations are valid:

$$r_m \cong \frac{u}{2}, \quad e \cong 1 \quad \text{and} \quad r_M = \frac{K_1 Z}{|E|} \tag{B.8}$$

We apply the quantization condition (1.43)

$$\oint_C m\bar{v}\, d\bar{s} = nh \quad \text{with} \quad n = 1, 2, \ldots \tag{B.9}$$

and obtain the following expression of the energy:

$$E = -R_\infty \frac{Z^2}{n^2} \quad \text{with} \quad R_\infty = \frac{m K_1^2}{2\hbar^2} = \frac{K_1}{2a_0} \tag{B.10}$$

where R_∞ is the Rydberg energy and a_0 is the first Bohr radius.

The period of the electron motion is

$$\tau = \frac{\pi n \hbar}{|E|} \tag{B.11}$$

The relations (B.6) become

$$a = \frac{n^2 a_0}{Z}, \quad b = \frac{n a_0 p_\theta}{Z\hbar}, \quad e = \sqrt{1 - \frac{p_\theta^2}{n^2 \hbar^2}}, \quad \text{and} \quad u = \frac{a_0 p_\theta^2}{Z\hbar^2} \tag{B.12}$$

In the quasipendular case, we have

$$r_M = \frac{2 a_0 n^2}{Z} \tag{B.13}$$

The average distance of the electron from the nucleus is

$$\tilde{r} = \frac{1}{\tau} \int_\tau r\, dt = \frac{n^2 a_0}{Z} \left(1 + \frac{e^2}{2} \right) \tag{B.14}$$

which, in the case of the quasipendular motion, becomes

$$\tilde{r} = \frac{3}{4} r_M \tag{B.15}$$

We calculate the equation of the Σ surfaces in the case of a hydrogenoid system, which is described in Section 1.6.2. Eq. (1.53), written in polar coordinates, r and θ, in the plane xy (see Figure 1.4), is (Landau and Lifschitz, 2000):

$$f(r_1, \theta_1) - f(r_0, \theta_0) = \int_{r_0}^{r_1} m \frac{dr}{dt} dr + \int_{\theta_0}^{\theta_1} p_\theta \, d\theta \tag{B.16}$$

There are two domains of integration of Eq. (B.16), as follows.

B.1 THE FIRST DOMAIN, WHEN r Increases from r_m and r_M

The point Q moves in the sense of increasing distance r, from the point A, which corresponds to $r_0 = r_m$ and $\theta_0 = 0$, to a current point of the C curve having the coordinates $r_1 = r$ and $\theta_0 = \theta$. Taking into account the initial condition (1.54), the solution of the integrals from Eq. (B.16) leads to the following expression:

$$f_1 = (\sqrt{\alpha r^2 + \beta r + \gamma})_{r_m}^r + \left(\frac{\beta}{2\sqrt{-\alpha}} \arccos \frac{2\alpha r + \beta}{\sqrt{\beta^2 - 4\alpha\gamma}} \right)_{r_m}^r$$

$$+ \left(\sqrt{-\gamma} \arccos \frac{\beta r + 2\gamma}{r\sqrt{\beta^2 - 4\alpha\gamma}} \right)_{r_m}^r + p_\theta \theta \tag{B.17}$$

We use normalized quantities. The action and angular momentum are normalized to $\hbar$ and the distances to a_0. The normalized quantities are underlined, as follows: $\underline{f_1} = f_1/\hbar$, $\underline{p}_\theta = p_\theta/\hbar$, $\underline{r} = r/2a_0$, $\underline{r}_M = r_M/2a_0$, $\underline{r}_m = r_m/2a_0$, and so on. With the aid of Eqs. (B.3)–(B.13), the expression of $\underline{f_1}$ results, as follows:

$$\underline{f_1} = \frac{2Z}{n} \sqrt{(\underline{r} - \underline{r}_m)(\underline{r}_M - \underline{r})} + n \arccos \left[\frac{1}{e} \left(1 - \frac{2Z}{n^2} \underline{r} \right) \right]$$

$$+ \underline{p}_\theta \arccos \left[\frac{1}{e} \left(1 - \frac{\underline{p}_\theta^2}{2Z} \cdot \frac{1}{\underline{r}} \right) \right] - \pi \underline{p}_\theta + \underline{p}_\theta \theta \tag{B.18}$$

It is easy to see that $\underline{f_1} = 0$ for $r = r_m$ and $\theta = 0$, and $\underline{f_1} = n\pi$ for $r = r_M$ and $\theta = \pi$.

With the aid of substitutions

$$\underline{r} = \sqrt{x^2 + \underline{y}^2} \quad \text{and} \quad \theta = \arccos \frac{x}{\sqrt{x^2 + \underline{y}^2}} \tag{B.19}$$

we write Eq. (B.18) in the form $f_1 = f_1(x, y)$. By virtue of Eq. (1.12), the equation of the de Broglie wave surfaces, corresponding to the upper part of the ellipse, is

$$\underline{f_1}(x, y) = \underline{\kappa} \tag{B.20}$$

where $\underline{\kappa}$ is a variable parameter.

B.2 THE SECOND DOMAIN, WHEN r Decreases from r_M and r_m

The point Q moves in the sense of decreasing distance r, from the point B, which corresponds to $r_0 = r_M$ and $\theta_0 = \pi$, to a current point of the C curve having the coordinates $r_1 = r$ and $\theta_0 = \theta$. The solution of the integral from Eq. (B.16) leads to the following expression:

$$f_2 = n\pi - (\sqrt{\alpha r^2 + \beta r + \gamma})^r_{r_M} - \left(\frac{\beta}{2\sqrt{-\alpha}} \arccos \frac{2\alpha r + \beta}{\sqrt{\beta^2 - 4\alpha\gamma}} \right)^r_{r_M}$$

$$- \left(\sqrt{-\gamma} \arccos \frac{\beta r + 2\gamma}{r\sqrt{\beta^2 - 4\alpha\gamma}} \right)^r_{r_M} + (p_\theta \theta)^\theta_\pi \tag{B.21}$$

By virtue of Eqs. (B.3)–(B.13), the relation of the normalized expression of f_2 results, as follows:

$$\underline{f_2} = 2n\pi - \frac{2Z}{n} \sqrt{(\underline{r} - \underline{r}_m)(\underline{r}_M - \underline{r})} - n \arccos \left[\frac{1}{e} \left(1 - \frac{2Z}{n^2} \underline{r} \right) \right]$$

$$- \underline{p}_\theta \arccos \left[\frac{1}{e} \left(1 - \frac{p_\theta^2}{2Z} \frac{1}{\underline{r}} \right) \right] + \underline{p}_\theta (\theta - \pi) \tag{B.22}$$

It is easy to see that $\underline{f_2} = n\pi$ for $r = r_M$ and $\theta = \pi$, and $\underline{f_2} = 2n\pi$ for $r = r_m$ and $\theta = 2\pi$.

With the aid of substitutions

$$\underline{r} = \sqrt{\underline{x}^2 + \underline{y}^2} \quad \text{and} \quad \theta = \pi + \arccos\frac{-\underline{x}}{\sqrt{\underline{x}^2 + \underline{y}^2}} \quad \text{(B.23)}$$

we write Eq. (B.22) in the form $f_2 = f_2(x, y)$, and the equation of the Σ wave surfaces, corresponding to the lower part of the ellipse, becomes

$$\underline{f_2}(\underline{x}, \underline{y}) = \underline{\kappa} \quad \text{(B.24)}$$

The equation of the C curve results, as follows:

$$b^2(x + a - r_m)^2 + a^2 y^2 - a^2 b^2 = 0 \quad \text{(B.25)}$$

The equation of the border of the CA domain, that is $E = U$, can be written as follows:

$$x^2 + y^2 = \frac{(2a_0)^2 n^4}{Z^2} \quad \text{(B.26)}$$

BIBLIOGRAPHY

Anderson, S.G., et al., 2004. Appl. Phys. B 78, 891–894.

Babzien, M.E., 2006. Phys. Rev. Lett. 96, 054802.

Corkum, P.B., 1993. Phys. Rev. Lett. 71, 1994–1997.

Coulson, C.A., 1961. Valence. Oxford University Press, London.

Courant, R., Hilbert, D., 1962. Methods of Mathematical Physics, vol. 2. Wiley-Interscience, New York, NY.

Courant, R., Lax, P.D., 1956. Proc. Nat. Acad. Sci. 42, 872–876.

de Broglie, L., 1924. Recherches Sur la Theorie des Quanta. Paris: Thesis.

de Broglie, L., 1956. Une Tentative d'interprétation Causale et Nonlinéare de la Mécanique Ondulatorie: La Théorie de la Double Solution. Gauthier-Villars, Paris.

de Broglie, L., 1966. Certitude et Incertitudes de la Science. Albin Michel, Paris.

Hartree, D.R., 1957. The Calculation of Atomic Structures. John Wiley, New York, NY.

Jackson, J.D., 1999. Classical Electrodynamics. John Wiley and Sons, New York, NY.

Kotaki, H., et al., 2000. Nucl. Instrum. Methods Phys. Res. A 455, 166–171.

Landau, L.D., Lifschitz, E.M., 1959. The Classical Theory of Fields. Pergamon Press, London.

Landau, L.D., Lifschitz, E.M., 1991. Quantum Mechanics. Pergamon Press, New York, NY.

Landau, L.D., Lifschitz, E.M., 2000. Mechanics. Butterworth Heinemann, Oxford.

Luis, A., 2003. Phys. Rev. A 67, 024102.

Messiah, A., 1961. Quantum Mechanics, vols. 1 and 2. North-Holland, Amsterdam.

Motz, L., 1962. Phys. Rev. 126, 378–382.

Motz, L., Selzer, L., 1964. Phys. Rev. 133, B1622–B1624.

Pogorelsky, I.V., et al., 2000. Phys. Rev. ST Accel. Beams 3, 090702.

Popa, A., 1996. Rev. Roum. Mathem. Pures Appl. 41, 109–117.

Popa, A., 1998a. Rev. Roum. Mathem. Pures Appl. 43, 415–424.

Popa, A., 1998b. J. Phys. Soc. Jpn. 67, 2645–2652.

Popa, A., 1999a. Rev. Roum. Math. Pures Appl. 44, 119–122.

Popa, A., 1999b. J. Phys. Soc. Jpn. 68, 763–770.

Popa, A., 1999c. J. Phys. Soc. Jpn. 68, 2923–2933.

Popa, A., 2000. An accurate wave model for conservative molecular systems. In: Kajzar, F., Agranovich, M.V. (Eds.), Multiphoton and Light Driven Multielectron Processes in Organics: New Phenomena, Materials and Applications. Kluwer Academic Publishers, Amsterdam, pp. 513–526.

Popa, A., 2003a. J. Phys. Condens. Matter. 15, L559–L564.

Popa, A., 2003b. J. Phys. A Math. Gen. 36, 7569–7578.

Popa, A., 2004. IEEE J. Quantum Electron. 40, 1519–1523.

Popa, A., 2005. J. Chem. Phys. 122, 244701.

Popa, A., 2007. IEEE J. Quantum Electron. 43, 1183–1187.

Popa, A., 2008a. Eur. Phys. J. D 49, 279–292.

Popa, A., 2008b. J. Phys. B At. Mol. Opt. Phys. 41, 015601.

Popa, A., 2009a. Eur. Phys. J. D 54, 575–583.

Popa, A., 2009b. J. Phys. B At. Mol. Opt. Phys. 42, 025601.

Popa, A., 2011a. Mol. Phys. 109, 575–588.

Popa, A., 2011b. Phys. Rev. A 84, 023824.

Popa, A., 2011c. Proc. Romanian Acad. A 12, 302–308.

Popa, A., 2012. Laser Part. Beams 30, 591–603.

Popa, A., Lazarescu, M., Dabu, R., Stratan, A., 1997. IEEE J. Quantum Electron. 33, 1474–1480.

Sakai, I., et al., 2003. Phys. Rev. ST Accel. Beams 6, 09101.

Slater, J.C., 1960. Quantum Theory of Atomic Structure, vols. 1 and 2. McGraw-Hill, New York, NY.

Slater, J.C., 1963. Quantum Theory of Molecules and Solids, vol. 1. McGraw-Hill, New York, NY.

Smirnov, V.I., 1984. Cours de Mathematiques Superieures, Tome IV, Deuxième Partie. Editions Mir, Moscow.

Synge, J.L., 1954. Geometrical Mechanics and de Broglie Waves. Cambridge University Press, Cambridge.

Wichmann, E.H., 1971. Quantum Physics, Berkeley Physics Course, vol. 4. McGraw-Hill, New York, NY.

Zauderer, E., 1983. Partial Differential Equations of Applied Mathematics. John Wiley and Sons, New York, NY.

Printed and bound by CPI Group (UK) Ltd, Croydon, CR0 4YY

07/10/2024

01041904-0004